DEVELOPMENTAL AND CELL BIOLOGY SERIES

EDITORS

P.W.BARLOW   P.B.GREEN   C.C.WYLIE

FROM EGG TO EMBRYO

FROM EGG TO EMBRYO

*DETERMINATIVE EVENTS IN EARLY
DEVELOPMENT*

J.M.W.SLACK

Senior Research Scientist, Imperial Cancer Research Fund

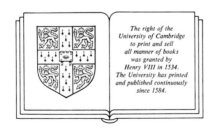

The right of the
University of Cambridge
to print and sell
all manner of books
was granted by
Henry VIII in 1534.
The University has printed
and published continuously
since 1584.

CAMBRIDGE UNIVERSITY PRESS

Cambridge

London New York New Rochelle

Melbourne Sydney

Published by the Press Syndicate of the University of Cambridge
The Pitt Building, Trumpington Street, Cambridge CB2 1RP
32 East 57th Street, New York, NY 10022, USA
296 Beaconsfield Parade, Middle Park, Melbourne 3206, Australia

First published 1983
First paperback edition 1984

Printed in Great Britain at the University Press, Cambridge

Library of Congress catalogue card number: 82–22168

British Library Cataloguing in Publication Data
Slack, J. M. W.
From egg to embryo. – (Developmental and cell
biology series; 13)
1. Embryology, Animals
I. Title   II. Series
591.1   QL955

ISBN 0 521 24205 3 hard covers
ISBN 0 521 27329 3 paperback

Contents

To Janet and Rebecca

Preface

This book is an enquiry into the mechanisms by which the spatial organisation of an animal emerges from a fertilised egg. It is intended for all students, teachers and research workers who are interested in embryos.

It is divided into three parts. The first two chapters introduce the problem of regional specification and attempt to define the meanings of embryological terms which are used in the remainder of the book. This is necessary because terms such as 'induction', 'regulation' or 'polarity' are often used but rarely defined and many controversies have arisen as a result of unnecessary misunderstandings.

The next four chapters give an overview of the experimental evidence which bears on the processes of cellular commitment from the time of fertilisation to the formation of the general body plan. The animal types considered are those on which most experimental work has been done: amphibians, insects, other selected invertebrates, the mouse and the chick. This is a general survey rather than a detailed review but sufficient references are provided to enable interested readers to pursue the topics in greater depth.

The last four chapters attempt to generalise the problems and to investigate the extent to which they have been solved by the theorists and the model-builders. In particular several 'gradient' models are examined and assessed in terms of the experimental evidence. This section has been written principally for non-mathematical readers although a few differential equations are provided for those who are interested. Many of the models are relevant to late development and to regeneration as well as to early development, but the focus of the book has been kept on early development because this provides for maximum conceptual unity without undue length.

I should like to thank the series editor, Chris Wylie, for the invitation to write the book; my wife Janet who expertly drew all the diagrams for Part II; Brenda Marriott for undertaking the lion's share of the typing; Richard Gardner, John Gerhart, Chris Graham, Brigid Hogan, Klaus

Sander, Jim Smith and Dennis Summerbell for commenting on portions of the manuscript; and finally the many authors who have provided me with copies of their photographs.

Part I

Introduction and definitions

1

Regional specification in animal development

This book is about how an egg becomes an animal. Attention will be concentrated on *early* development because this is the time at which the important events are happening. As everyone knows the human gestation period is about 9 months long, but it is not so commonly appreciated that the basic body plan of the embryo becomes established during the very short period from 1 to 4 weeks after fertilisation. During this time an apparently homogeneous group of cells, the inner cell mass of the blastocyst, becomes transformed into a miniature animal consisting of central nervous system, notochord, lateral mesoderm, somites, branchial arches, integument and gut. All of these parts contain specific types of cell and all lie in the correct positions relative to one another. In later development there is a good deal of growth and of histological differentiation of the organs, and the specifically human, rather than the general vertebrate, characteristics of the organism become established. However, all this takes place on the framework of the basic body plan which was laid down in early development. As Wolpert has emphasised, it is not birth, marriage or death, but gastrulation which is truly the most important time in your life.

The core problem of early development is that of *regional specification*, also called *pattern formation* or *spatial organisation*. This refers to the process whereby cells in different regions of the embryo become switched onto different pathways of development. It is the mechanism of this process with which we are primarily concerned in this book. Regional specification should not be confused either with *cell differentiation* or with *cell movement*, which are processes posing us with important but distinct problems. Cell differentiation nowadays means the synthesis of new species of protein – species which are different from those made by the cell itself or by its ancestors at previous times, and different from those made by other cells in the embryo at the same time. The movement of cells in the embryo is by no means understood but is presumably related to cell differentiation in the sense that a cell expressing a certain group of substances on its surface will interact with its neighbours and with the extracellular matrix in such a way as to migrate in a certain direction and

3

stop at a certain place. These processes are of immense importance and clearly without them there would be no development. However, they are not extensively discussed here because they are consequences and not causes of regional specification. Of course once the first regional distinctions have been made the competence of cell populations to make further decisions will depend on the specific differentiation already achieved, but it is still possible to distinguish between the process of specification and its consequences.

The book itself is divided into three sections. Chapter 2 surveys the concepts of experimental embryology in an attempt to establish an unambiguous language for discussing the phenomena. Chapters 3 to 6 review what is known about regional specification in a variety of types of animal embryo. Chapters 7 to 10 ask what sort of mechanisms have been proposed to account for the phenomena and how well they stand up in the light of the evidence.

Universality and homology

For ethical reasons most experimental embryology relates to animals other than the human. Mammalian embryos are usually felt to be the best models for man but viviparity poses some serious technical difficulties and so most experimental embryology before the recent period has concerned itself with invertebrates and non-mammalian vertebrates. It is therefore an issue of some importance to know how similar are the mechanisms for regional specification in different types of animal. There are two ways of looking at this. On the one hand the universalist will say that the mechanism of inheritance, protein synthesis and cell ultrastructure are the same in all eukaryotic organisms so probably the mechanisms of early developmental decisions will be the same as well. On the other hand his opponent would argue that it is the formation of the basic body plan which is at stake here and different animal phyla are distinguished from one another precisely because they have different basic body plans, so we might expect their mechanisms of formation to be different.

Where different animals contain visible homologous parts which arise in homologous ways, it seems probable that the same biochemical mechanisms underly their formation. For example the neural tube arises from dorsal ectoderm in all vertebrates. In amphibians and chicks there is good evidence that it is formed as a result of a signal from the underlying mesoderm, and there is a little evidence that the same is also true in ascidians. It would be surprising if different mechanisms were at work in a case like this where not only is the final morphology the same but similar cellular interactions bring this morphology into being.

A similar argument can be made for the different species of insect, whose embryos can be seen to have the same basic structure at the late germ band stage. Double posterior duplications (double abdomens) can arise in the fruit fly, in the midge and in the leafhopper as a result of various manipulations (see Chapter 4). The fact that the same striking abnormality in the body plan can arise in different species following experimental interference again argues for the existence of a common underlying mechanism.

But is there anything in common between insects and vertebrates? The basic body plans are completely different and the observable course of early development is also completely different. If there is a common mechanism for early developmental decisions it will presumably be one which was already in existence in the common metazoan ancestor of vertebrates and insects, and we do not know anything at all about such an ancestor, or even that there ever was one.

On the other hand there are a number of properties of the early developmental mechanisms which do seem to be similar not only between vertebrates and insects but between all the groups of animals considered in this book. These are not related to particular anatomical features but rather to behaviours in the face of experimental interference, one example being the universal ability of early embryos to regulate their proportions following an alteration in overall size. These properties are collected and discussed in Chapter 7, but whether the similarities are significant or merely fortuitous will not be known for certain until the mechanisms themselves are understood at a biochemical level. Perhaps the problem of universality is itself a deep enough question to justify the study of types of embryo far removed from man, although the exigencies of funding for medical research will inevitably continue to favour the vertebrates at the expense of the invertebrate phyla.

The developmental hierarchy

It is very important to emphasise that even the basic body plan is not specified all at once but is formed as a result of a hierarchy of developmental decisions. This statement will be justified more fully by the experimental results reviewed in Part II, but for the moment consider an embryological lineage such as that illustrated in Fig. 1.1. This shows the provenance of different regions of the vertebrate body deduced mainly from the experimental embryology of the Amphibia. The diagram comprises a number of subdivisions of different multicellular regions. Without considering the mechanism of these decisions, which is the subject of the remainder of the book, we note that the familiar histological cell types are to be found only on the right side of the diagram. Each decision is irreversible and so all the states which precede terminal differentiation each embody a reduction of potency compared with the previous state. For example, for cells to be competent to form the lens of the eye they

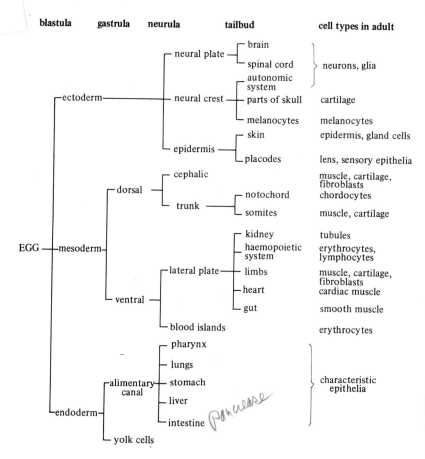

Fig. 1.1. Formation of the basic body plan in a vertebrate (excluding extra-embryonic regions). By the early tailbud stage the embryo consists of a mosaic of regions determined to form the principal organs and structures of the body. This body plan is built up as a result of a hierarchy of decisions, and several further decisions will in most cases be taken before the cells differentiate into the terminal cell types shown on the right-hand side. It should be noted that some cell types, such as cartilage, arise from more than one lineage. Further details will be found in Chapters 3 and 6.

must have already decided that they are ectoderm (rather than mesoderm or endoderm), and epidermis (rather than neural plate or neural crest). So each decision is made by cells in a particular determined state, it is made among a small number of alternatives and the outcomes are not in general terminally differentiated cell types but new states of determination whose potency is further restricted. It is because of the conviction that the basic body plan arises from a *hierarchy* of decisions between determined states that this book is subtitled 'determinative events in early development'. Unfortunately, with the exception of the three 'germ layers', whose existence was deduced in the nineteenth century from descriptive embryology, there are no suitable names for the determined states. We are thus driven to referring either to positions ('anterior mesoderm', 'posterior mesoderm') or to the future organ ('limb field', 'eye field'), and this makes embryological terminology rather obscure and confusing to the outsider.

Only two general processes are known which account for the events of regional subdivision: *cytoplasmic localisation* and *induction*. We do not understand either very well and in no case do we know the biochemical basis of the phenomena. Cytoplasmic localisation has been studied mainly in invertebrate embryos with a small total cell number (see Chapter 5). It appears that regulatory molecules, whose identity is unknown, become differentially distributed in the cytoplasm of a stem cell. In the course of an asymmetrical cell division the two daughter cells inherit different materials and consequently enter different states of determination.

Induction has been studied mainly in vertebrate and insect embryos. Here it appears that the competent tissue becomes differentially determined in response to chemical signals from other regions of the embryo. The possible mechanisms for induction are extensively discussed in Chapters 9 and 10.

The distinction between cytoplasmic localisation and induction is one which has been made using classical techniques of experimental embryology and it is clear that both processes are widespread and probably occur in all types of embryo. Unfortunately the classical techniques are limited when it comes to analysing further the mechanisms of these processes and this becomes particularly obvious when we consider the near impossibility of excluding various theoretical models which have been proposed in recent years. Obviously new experimental methods are needed, but exactly what they should be is not so clear.

Positional information and non-equivalence

The recent upsurge of interest in regional specification owes a good deal to the 'positional information' concept of Wolpert (1969, 1971). This is often confused with a specific model for induction, the diffusible morphogen model, which is discussed in Chapter 9. However, the doctrine does not involve belief in any particular model but rather a set of general precepts about embryonic development. The term 'positional information' has itself become rather ambiguous and is not used here, but the general precepts are listed since they have influenced the author in terms of the material selected for inclusion and the manner of its presentation:

1. Regional specification comes first, cell differentiation and cell movement are consequences.
2. Regional specification can always be broken down into two independent processes: an instructive process during which positional information is imparted, and an initial response by the competent tissue called 'interpretation'.
3. The biochemical mechanism underlying positional information is the same in all animals. The mechanism of interpretation differs according to the particular anatomy being formed.
4. Cells which end up with the same histological type, but which are of different embryological provenance, are at least transiently 'non-equivalent', i.e., exist in different states of determination.

Wolpertians will be able to read this book if they substitute 'positional signal' for 'inductive signal' and 'positional value' for 'state of determination'. The more traditional terms have been employed here both because they are more generally understood and because the Wolpertian terminology does not lend itself to situations involving a hierarchy of decisions which are typical of early embryos.

The concept of non-equivalence (Lewis & Wolpert, 1976) is particularly important because it is implicit in most modern thinking about regional specification and has far-reaching implications. The idea is that in an organ such as the limb, which consists of quasi-repetitive arrangements of muscles and skeletal elements, the different parts of the embryonic rudiment necessarily acquire *different* states of determination as a result of the regionalisation process, even though they may end up by becoming the same histological cell type. In connection with this it should be noted that the same cell type can arise by different routes in different parts of the embryo (Fig. 1.1). Evidence for non-equivalence can be found in the fact that the program of maturation may be visibly different: for example cartilages of the wrist grow less than those of the long bones (Summerbell, 1976) and different cartilages become ossified at different times (Holder, 1978). In cases of animals which are capable of extensive regeneration it seems probable that the information encoded in the states of determination persists into adult life and can be recalled following the removal of parts (Slack, 1980*a*).

The implications of non-equivalence are twofold. First, if this way of looking at things is correct, the process of regional specification is complex. As we shall see in Chapter 10, it can be shown that an organism consisting of n parts will need something more than n biochemical components (genes, switches, substances) to specify that they should develop in the correct relative positions. So an organism as complex as man would have an appreciable fraction of the genome devoted to controlling regional specification. This is an area of cellular activity about which we presently know next to nothing at the biochemical level, and we can anticipate that its unravelling may take several decades of research activity by biochemists and molecular biologists.

Secondly, if non-equivalence is not transient but persists into adult life, then the organism must possess a 'second anatomy' of codings expressing embryological decisions, which is superimposed on the familiar visible anatomy. The implications of this are discussed elsewhere (Slack, 1982).

Since positional information and non-equivalence are inseparable concepts it follows that mechanisms which generate a series of structures in identical states of determination are not positional information mechanisms. Some of these will be discussed in Chapter 10 in connection with the determination of repeating structures such as segments and somites.

Molecular biology

The other major new influence on contemporary thinking in embryology comes from molecular biology and is, for example, well articulated by Davidson (1976). Molecular biologists argue that different sorts of cells exist because they are expressing different genes and therefore the control of gene expression is the central problem of developmental biology. Some progress has been made in recent years in identifying regulatory sequences which control gene expression in animal cells and as this is such an active area of research it is sure that the molecular details will soon be known in a few cases.

Unfortunately, for the embryologist it will not help very much to know how one or two particular genes are regulated. To understand the course of development it will be necessary to predict the pattern of gene activity in a cell from knowing the previous pattern of activity and the significant influences from the environment. This cannot be done without knowing what controls the activity of each and every gene. In bacteria we know that different genes are regulated by different substances and indeed that there are several basically different types of mechanism involved. The number of genes in animals is not really known but is usually estimated in the range 10^4–10^6 (see Chapter 8), so it will certainly take a great deal of time to build up enough knowledge of gene regulation to be able to explain, or predict, the course of embryonic development.

Despite rapid progress in molecular biology it must be admitted that it has yet to make a major contribution to our understanding of early development. This may be because some of the gene products involved in terminal differentiation are known while those involved in early determinative decisions are not. In the following chapters biochemical and genetic results are included and discussed where they are relevant, but the fact is that most of what we know has come from microsurgical experiments and this is reflected in the coverage given. It is certain that biochemical investigations will become more significant in the future although an exclusive concentration on gene expression may not be entirely appropriate since the events of regional specification in all probability do not involve gene expression except as a rather late consequence. To take just one example; in the case of the expression of acetylcholinesterase in the ascidian tail muscle, which will be discussed in Chapter 5, the *cells* become specified by the 8-cell stage but the *nuclei* only become specified by the 76-cell stage, and *transcription* does not commence until one or two more cell divisions have occurred after that.

However, it is to be hoped that molecular biologists will find something in this book to interest them. Most of the progress in their science has been achieved using bacteria, viruses and cultured cells, but if they are really interested in the mechanism of development they will have eventually to study the embryo itself, and this means in the first instance being familiar with the data that already exist and the models which have been advanced to explain them.

2

The concepts of experimental embryology

Since the late nineteenth century a distinction has been drawn between 'descriptive' and 'experimental' embryology. *Descriptive* embryology refers to an account of the course of development based either on the view down a dissecting microscope or on reconstructions of the successive embryonic stages from serial histological sections. *Experimental* embryology seeks to discover the mechanisms which underlie the visible changes and to do so by perturbation of the normal course of development.

This distinction is probably not now as valid as it used to be. Morphological description does throw some light on mechanism because if often allows us to eliminate some model mechanisms straight away. Likewise many experiments are carried out in such a way that they do not disturb the normal course of development, particularly those concerned with fate mapping or the discovery of biochemical markers for different stages or regions of the embryo. However, historically it is the experimental embryologists who have given most thought to mechanism and who have formulated the basic conceptual framework which enables us to discuss the problems rationally. In this chapter an attempt will be made to present at least a vocabulary which can serve as the currency for the arguments in the remainder of the book. Although an attempt has been made to conform as much as possible to common usage, the exigencies of precise definition ensure that some readers will find words used in different senses from those which they favour.

Normal development and the fate map

Normal development means the course of development which a typical embryo follows when it is free from experimental disturbance. It must not be confused with pathways of development which give a normal outcome. For example, when the first two blastomeres of an embryo are separated and each develops into a complete animal the outcome may be normal but the pathway is not normal with respect either to the regions of the egg which become each part of the adult or to the absolute dimensions of parts before compensatory growth has occurred.

A *fate map* is a diagram which shows what will become of each region of the embryo in the course of normal development: where it will move, how it will change shape, and what structures it will turn into. The fate map will change from stage to stage because of morphogenetic movements and growth, and so a series of fate maps will depict the trajectory of each volume element from the fertilised egg to the adult. While it is clear that a series of fate maps must exist for the stages of each individual embryo it is only worth knowing a fate map if it is the same for different individuals and therefore has predictive value. It is in this sense that we say a fate map 'exists' for a certain type of embryo. What we mean is that there is no random mixing of cells for the stages in question, for if there is some mixing then a cell in a certain position in one embryo may become something quite different from the cell in the same position in another embryo.

In the complete absence of mixing a fate map can, in principle, have unlimited precision. Not only will there exist a family tree showing the lineage of every cell in the adult body, but each volume element of the original egg cytoplasm will end up in predictable cells of the adult. This situation is approached by some invertebrates in which the adult cell number is relatively small, and we shall return to them in Chapter 5. For most embryos, however, there is reason to think that there is usually a little local mixing of similar cells and therefore the fate maps cannot be quite this precise. Nonetheless the fate map is an absolutely fundamental concept in embryology and the interpretation of nearly all experiments concerned with early developmental decisions depends on knowledge of the fate map.

In practice it is rarely possible to be confident about the fate map from histological observations alone and so some means has to be found of marking specific regions of the embryo. So long as the marking procedure is gentle enough not to disturb normal development then a mark applied at time t_1 can be located at time t_2 and we can deduce retrospectively that the t_1 mark was the prospective region for the t_2 mark (Fig. 2.1). It is essential to note that prospective regions are not necessarily *committed* to become the structures which they later form in normal development. An ill-founded belief that they are has caused endless confusion in the past and has sometimes, unfortunately, led to the term 'fate map' being used for maps of potency or states of determination – distinct concepts which will be discussed below.

If there is no mixing then the boundaries of prospective regions will abut one another as shown in Fig. 2.2(*a*). If there is a little mixing the boundaries will overlap because they include all the cells that have some probability of being incorporated into the structure concerned. The prospective regions on the map are thus larger than the actual prospective regions in particular individuals (Fig. 2.2*b*). In the extreme case where

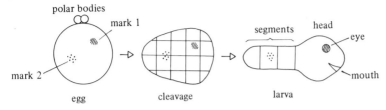

Fig. 2.1. The principle of fate mapping. In the absence of random cell mixing, marks placed on the egg will label particular regions of the larva, in this case the eye and the second abdominal segment.

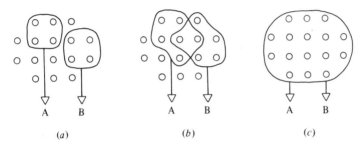

Fig. 2.2. Dependence of the 'primordial cell number' on the amount of mixing which occurs after marking. (*a*) If there is no mixing at all then the prospective regions are the same size as the later structures A and B. (*b*) If there is some mixing then the prospective regions are larger than the later structures because they include all cells which have some probability of contributing to them. (*c*) If there is complete randomisation of position then any cell can contribute to any structure.

there is total randomisation of position there is no fate map at all since any cell, or region of the egg, has a finite probability of ending up in any structure whatever (Fig. 2.2*c*).

A variety of experimental techniques has been used to construct fate maps and some of the results will be described in Part II. An ideal marker is one which can be applied at will to any region, at any stage, which is readily visible at all later stages, and which does not perturb normal development. Needless to say no such marker exists and all methods are to some extent compromises between these qualities. Some eggs possess visible regional differences in their cytoplasm which can be traced through to some of the daughter cells (e.g. Conklin, 1905*a*). These are useful but leave the rest of the embryo unlabelled. 'Vital stains' are dyes which can be applied to parts of embryos without damaging them (e.g. Vogt, 1929). They have been extensively used in amphibian and chick embryology but have limited resolution because of their tendency to spread and fade. Grafts of tissue labelled with tritiated thymidine are useful if the label is not diluted too much by cell division (e.g. Nicolet, 1971). The labelled nuclei can be visualised in histological sections by autoradiography. In mammalian and insect embryology, genetic markers have been popular, usually mutations affecting pigmentation or non-essential enzymes (e.g. Gardner, 1978; Janning, 1978). The main problem with them is that they are rarely genuinely cell-autonomous. Either they are only expressed in some cell types, or their expression is sensitive to a cell's environment, or both. Cytological markers, particularly those involving nucleolar number, are almost completely cell-autonomous but may be difficult to visualise in sections. Markers in interspecies combinations such as the chick–quail or rat–mouse systems are only suitable for special applications since the cells of different species are usually too different to undergo exactly normal development in combination with one another.

Growth

It is often assumed that growth is an essential aspect of embryonic development, but this is not so. Of course the cell number will always increase, but cell divisions in early embryonic development are typically cleavage divisions which produce daughter cells half the size of the mother cell. Only viviparous animals are truly in a position to grow in the sense of putting on dry weight since they possess a food supply external to the egg. The majority of animals are not viviparous and lay eggs which develop in isolation. It is frequently the case that the egg develops an embryonic and an extra-embryonic region, the latter containing a lot of yolk, and in such cases the embryo will grow at the expense of the yolk mass. But there are also many animals whose eggs contain little yolk and which do not increase in dry weight until they have hatched and can feed themselves.

The occurrence of growth during embryonic development reduces the scope of the fate map to some extent since it means that cytoplasmic regions cannot be traced from the egg to the later embryo. However, the fate map will still exist at the level of the cell nucleus: for example a cell marked with a somatic mutation will produce a clone which occupies a certain region of the embryo and can later be identified by its distinctive phenotype. The yolk itself is not included in the fate map since it is assumed that any part of it may furnish the matter to construct any part of the embryo.

Stem cells

Stem cells are cells which undergo unequal cell divisions to produce dissimilar daughters (Fig. 2.3c and d), and can be contrasted to ordinary

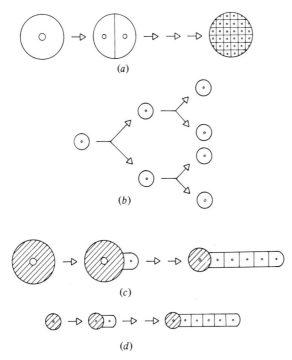

Fig. 2.3. Types of cell division found in embryos. (*a*) Cleavage divisions unaccompanied by growth. (*b*) Exponential proliferation supported by an external nutrient supply such as a yolk mass or a placenta. (*c*) and (*d*) Unequal divisions of a stem cell produce a string of progeny which are not themselves stem cells. In (*c*) there is no growth and in (*d*) there is growth.

cells which produce two similar daughters on division (Fig. 2.3*a* and *b*). Stem cells can sometimes be seen at work in invertebrate embryos such as the leech (Stent & Weisblat, 1982). They are not normally thought to be important in the early development of vertebrate or insect embryos, but it may be that they have simply not been looked for very hard and it is possible that the growth zones which produce the vertebrate tail or the insect abdomen actually contain stem cells. In fate map terms a stem cell (or growth zone) is the prospective region for *all* of its progeny. Any regional organisation which is produced in the program of cell divisions from a stem cell will not be represented on the fate map of earlier stages, so those who think that stem cells are very important tend not to be interested in fate maps.

Stem cells probably work by developing a cytoplasmic asymmetry before their division, such that regulatory molecules inherited by the two daughter cells evoke different patterns of gene activity in their nuclei. It is also thought possible that the individual cells in the sequence of progeny produced from a stem cell may differ from one another as a result of some sequential chromosomal modifications in the parent cell, but there is no real evidence for this up to the time of writing.

In cancer research the term 'stem cell' tends to be used more loosely to mean any cell capable of indefinite multiplication; this definition will not be used here since all cells in an early embryo will have some progeny capable of indefinite multiplication.

Clonal analysis

Clonal analysis is a form of fate mapping in which a single cell is marked and the position and cell types of its progeny identified at a later stage. The principles of fate mapping apply in so far as the marked cell is judged to be the prospective region for its own progeny. Its special interest lies in the deductions which it enables us to make about determination. The full complexities of 'determination' will be discussed below, but for the moment we can accept that a determined cell is one that is irreversibly committed to develop into part of a given structure. Now since a single cell may be determined to form one of two structures, say A and B, or neither, but not both, it follows that if a clone overlaps both structures, the marked cell cannot have been determined at the time it was marked. In other words *no clonal restriction means no determination*. The negative form of this principle is crucial because the converse is not true. Clonal restriction by no means implies determination; it simply means that the marked cell is not in the prospective region of the structure to which it does not contribute (Fig. 2.4). The frequent failure to understand this is due entirely to a failure to understand exactly what is meant by a fate map, a difficulty which should not now be shared by readers of this book.

The power of clonal analysis is greater the more cellular mixing takes place before determination. This is because mixing increases the size of the prospective regions and makes it more likely that a given clone will be in more than one prospective region. The clone must of course have become at least two cells by the time of determination if it is to have any chance at all of spanning the boundary. After determination there can still be random mixing within each territory but not across the boundary since by definition the cells are irreversibly committed and so any transgression of the boundary would disturb the final anatomy.

The use of clonal analysis can enable us to exclude a possible stem cell mechanism for producing a structure. If marked clones are induced at random at a stage before a stem cell starts dividing then the cell must either belong to a clone or not, and the resulting structure will accordingly either be completely marked or completely unmarked. If the structure is partially marked then it must be formed from at least two of the cells present in the embryo at the time of marking.

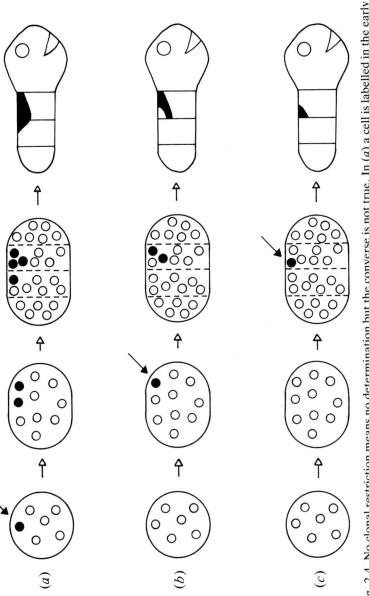

Fig. 2.4. No clonal restriction means no determination but the converse is not true. In (a) a cell is labelled in the early embryo and its progeny span the boundary between two segments in the adult. In (b) a cell is labelled before determination but because of its position its progeny do not span the segment boundary. In (c) a cell is labelled after determination and its progeny are also confined to one segment.

Other things being equal, clones arising from cells marked early in development will end up larger than those induced later because they undergo more cell division. This fact is sometimes used to estimate 'primordial cell numbers' on the grounds that if there is a total of n cells in the prospective region at the time of marking, then the marked clone will subsequently occupy about $1/n$ of the structure. This type of calculation assumes that there is no differential growth, either of different parts of the structure or arising from the marking itself. The procedure is valid if clones are induced after determination since the determined rudiment will possess a countable number of cells at each stage, but if they are induced before determination there is a difficulty. When the late embryo is examined the clone will either be confined to the structure in question or it may overlap other structures. If it is confined then its size will underestimate the size of the prospective region on the fate map since some of the other cells of the prospective region will have contributed to different structures. If the clone is not confined then the proportion of the structure labelled will lead to an overestimate of the size of the prospective region since some of the marked cell's progeny have gone elsewhere. Estimates of primordial cell numbers therefore need to be treated with some care.

Forms of developmental commitment

We have seen that a fate map conveys no information about the intrinsic character or the commitment of the prospective regions. At present our understanding of the biochemistry of early embryos is too primitive to allow us to learn about commitment from this source and so our definitions are operational ones based on particular types of experiment.

In this book, three types of commitment will be distinguished: specification, determination and potency. A tissue explant is *specified* to become a particular structure if it will develop autonomously into that structure after isolation from the embryo. The medium into which the explant is isolated must be 'neutral' with respect to the developmental pathway followed. The nearest approach to this situation is found in amphibian embryos where tissue explants will develop and differentiate *in vitro* in buffered salt solutions, and in a number of invertebrate embryos where it is possible to obtain quite advanced development of parts of the body from isolated blastomeres. The specification of a region need not be the same as its fate in normal development. For example the prospective neural plate of an amphibian blastula will differentiate not into a neuroepithelium but into epidermis when cultured in isolation (Holtfreter, 1938*a*, *b*). In principle it is possible to construct a 'specification map' of the embryo by explanting tissue from each region and combining the results. As we shall see later, if the specification map is the same as the fate map we refer to the region concerned as a 'mosaic'. If the specification map differs from the fate map we refer to the system as 'regulative'.

A *determined* region of tissue will also develop autonomously in isolation but differs in that its commitment is irreversible with respect to the range of environments present in the embryo. In other words it will continue to develop autonomously after grafting to any other region of the embryo. A very large number of embryological experiments basically consist of grafting a piece of tissue from one place to another and asking whether it develops in accordance with its new position or its old position. In the former case it is not determined, although it may have been specified, and in the latter case it is determined. A series of such grafts performed at different stages usually show a time at which the tissue becomes determined. For example, the prospective neural plate of an amphibian embryo becomes determined to form neural plate at gastrulation. After this stage grafts of this tissue to other regions of the embryo will always form neuroepithelium. At the earlier blastula stage it is not determined. It will form neural plate if left *in situ*, but if grafted ventrally it will form epidermis or if grafted vegetally it will form mesodermal structures (Spemann, 1938).

The *potency* of a region of tissue is the total of all the things into which it can develop if put in the appropriate environment. This is not easy to measure experimentally because one can never be sure that all possible environments have been tested. In particular the range of environments present in the embryo at the stage in question may not be sufficient to enable the tissue to access all of the possible states. For this reason the potency of a tissue is not the same thing as its state of determination but may be wider. Conversely the 'restriction of potency' is also not the same thing as determination, although it is nowadays often defined in this way by mammalian embryologists. If the tissue can form more than one thing it is called 'pluripotent' and if it can form everything it is called 'totipotent'. Of course a totipotent tissue cannot necessarily form a complete embryo in isolation because the appropriate range of environments which it requires for its regional subdivision may not be present.

Specification, determination and potency can be defined only for the particular level of the hierarchy of decisions corresponding to the developmental stage in question and this can cause difficulty when the observables are the product of subsequent decisions. For example, the first decision in Fig. 1.1 is between the three germ layers, but we can only know that a region of tissue is committed to form mesoderm if it later produces characteristic mesodermal cell types such as muscle, mesenchyme, kidney, etc. Which of these are produced and how they are arranged will depend on the ability of the graft or explant to generate internal regional subdivisions, as well as on its overall state of commitment. These later decisions may of course be altered independently of the state of commitment, and so, for example, the experimenter can be left unsure of whether the tissue was originally committed to form mesoderm, or somite, or muscle, each of these states being a different level in the hierarchy of decisions. The way out of this difficulty is to find biochemical markers characteristic of the states in question rather than of their subsequent derivatives. This has been done to a greater extent in mouse embryology than in other fields up to the present time (see Chapter 6), although even here the situation remains unsatisfactory.

The above definitions have been given in terms of regions of tissue. They can also be applied to single cells or even to nuclei. When a cell is said to be determined this carries with it the implication that the state is clonally inherited, since it must be independent of environment. So a cell can acquire a particular state of determination either because of its lineage or, *de novo*, by responding to appropriate external signals. In fact the states of determination of nuclei, single cells and tissues are often not the same. As we shall see, it has been difficult to demonstrate any determination at all of nuclei. Single cells may appear less determined than the tissues from which they come simply because the environment in the centre of a graft is mainly the product of the graft cells themselves, and it may indeed be the case that some states of determination require some particular three-dimensional arrangement of cells, or some extracellular materials, and simply cannot be maintained at the single-cell level.

Polarity

The term polarity is commonly used in two senses when applied to tissues. The most common meaning, and that which will be adhered to below, is regional difference in state of commitment. So when we refer to 'determination of dorsoventral polarity' we mean the establishment of two or more distinct states of determination along the line from the dorsal to the ventral extremum of the embryo.

But polarity is often applied to single cells to indicate that they are visibly different at the two ends: for example the basal and apical surfaces of epithelial cells. It is therefore possible to describe a tissue as polarised if it is composed of polar cells with a preferred orientation, even though there is no regional heterogeneity.

Regulation and mosaicism

Regulation and mosaicism are old embryological terms and are defined here in such a way that they relate most directly to the presumed mechanisms of regional specification. Mosaic behaviour then implies that cells do what they do in normal development while regulative behaviour implies that they do something different. For this reason, regulation and mosaicism should only be defined for situations in which the fate map is known. It should be emphasised that both types of behaviour can occur in a given type of embryo, an early regulative phase usually preceding a later mosaic phase.

Mosaicism is fairly straightforward. An embryo, or part of an embryo, is mosaic when the map of specified regions coincides exactly with the fate map. So if it is cut into pieces, each part develops in the same way as it would in normal development. There is usually a mosaic stage which immediately precedes overt differentiation, and in such cases the significance of this mosaicism is no greater than that of the final anatomy since differentiation may already be under way but not yet visible down the microscope. In fact modern biochemical methods often make it possible to detect the onset of differentiation earlier than does light microscopy (e.g. Rutter *et al.*, 1968). Mosaicism is more interesting when it is detected in very early embryos, as will be discussed in Chapter 5, because this is clearly long before the onset of terminal differentiation. This is often regarded as evidence for the existence of cytoplasmic determinants in the egg which control the specification of parts, although it is in fact only consistent with this possibility and does not by any means prove it.

In general, *regulation* can be defined as the complement of mosaicism, so that an embryo or organ rudiment is regulative if the state of specification of its isolated parts does not correspond to the fate map. This idea is, however, too vague for practical purposes and it is necessary for us to consider several types of regulative behaviour since each has somewhat different implications for the nature of the underlying mechanisms. These are twinning, fusion, defect regulation and inductive reprogramming.

Twinning means the production of two complete animals from an embryo divided into two parts. As we shall see, this is quite a common though not universal property of embryos at the 2-cell stage. This is true, for example, of the mouse, the frog, the sea urchin and the starfish. Complete twinning from more advanced stages containing thousands of cells can occur in the camel cricket, the duck, the newt and, probably, humans. In all these cases the principal body axes are half normal size at the time of formation. In embryos capable of growth the size of each twin approaches the normal during foetal or larval life. The proportions of parts within the miniature primary axes are, however, normal at all times.

The phenomenon of twinning reveals several features of normal development which are somewhat obscured by the simple designation of the embryos as 'regulative'. First, it excludes any model for regional specification which is based solely on the localisation of determinants in the egg cytoplasm. Any such determinants have to be shared out in such a way that they occupy the same relative positions in the two fragments and from simple geometrical considerations this means that they must be symmetrically disposed with respect to the plane of division (Fig. 2.5). It

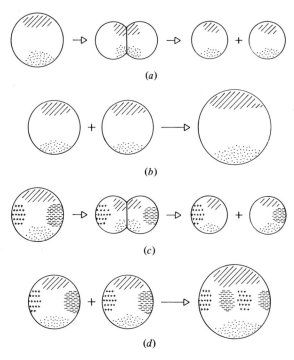

Fig. 2.5. Embryonic regulation and cytoplasmic determinants. (*a*) and (*c*) show blastomere isolation experiments and (*b*) and (*d*) show the fusion of two eggs to form a giant embryo. In either case individuals with a normally proportioned anatomy can only arise if the determinants are symmetrically disposed along the plane of separation or joining. In (*a*) and (*b*) this is the case and in (*c*) and (*d*) it is not.

therefore implies that most of the regional subdivision arises after fertilisation as a result of interactions between different parts of the embryo. Secondly, it shows that these interactions are able to accommodate a change of scale in the pattern. Boundaries which would be formed 100 micrometres apart in the normal embryo will be formed 100/cube root of 2 = 79 micrometres apart in the twins. This is quite an important constraint on possible signalling mechanisms, and we shall return to it in some detail in Chapter 8. Thirdly, it shows that the final size of each structure does not depend on the size of the original primordium but on some mechanism which can stop growth when a certain absolute size has been reached. This is a problem of growth control in late development and adult life and is therefore outside the scope of this book, although it is no less significant for that.

The property of forming normally proportioned half-size patterns is also possessed by some organ rudiments; for example, this is true of the eye rudiment in amphibian neurulae (Adelman, 1930) and the limb rudiment in amphibian tailbud stages (Slack, 1980*b*). The implications for the mechanism of regional specification are the same as in the whole embryo except that for an organ rudiment there is no need to assume any subdivision at all at early stages since the signals which eventually bring it about may be emitted by the surrounding tissues.

The converse of twinning is fusion of two or more embryos to give one giant embryo. This has similar implications for mechanism because any cytoplasmic determinants have to be symmetrically disposed with regard to the fusion plane. In the fused embryo the parts of the body plan exceed normal size at the time of formation and growth is relatively slowed during late development to bring them back to normal size. This phenomenon has been extensively investigated in mouse embryos where production of tetraparental mice by fusion of morulae has become a routine procedure. It is further discussed in Chapter 6.

Many embryos are not capable of producing twins, but there are few if any which are not capable of some degree of defect regulation. This means that there is at least some region of the embryo which can be removed at an early stage without disturbing the pattern or the proportions of the body plan. Defect regulation tells us about the regions of the embryo which must be present in order to obtain a complete pattern. For example in the sea urchin morula, quite a lot of tissue can be removed from the equatorial region without resulting in the production of defects while only a little can be removed from the terminal regions. In the amphibian blastula ventral tissue can be removed but dorsal tissue cannot. It is natural to identify the regions whose presence is obligatory with the signalling centres or organisers of the early embryo. We do not know in any case what these centres actually do but some possibilities are mentioned in Part III.

A fourth type of phenomenon which should be called regulation, but often is not, is inductive reprogramming. If a signalling centre is grafted to an abnormal position it can cause the surrounding tissues to follow a pathway of development which does not correspond to the fate map. This is often not regarded as regulation because the outcome is not a normal embryo, but at the cellular level the processes are probably the same as those following fragmentation or fusion.

Regulation has been discussed here as though it was always perfect. This is not in fact the case; more often than not it is probably imperfect. For example, twins may sometimes be lopsided, the side of origin being bigger than the regulated side. There are few careful studies of proportions in regulated embryos and those that there are often reveal imperfections in embryos which are superficially normal (Marcus, 1979). However, this is not of great significance. So long as there is some adjustment of the fate map in response to the operation the conclusions reached here hold good. We do not know at present whether a failure to regulate completely is an intrinsic inability of the mechanisms to do so or whether it is simply caused by lack of time before the secondary embryonic interactions commence.

Finally, it is important not to confuse embryonic regulation as discussed here with *regeneration* in adult or larval animals. Regeneration involves the re-establishment of regional differentiations or determinations which have already been set up, while regulation involves the re-establishment of a fate map on a partial domain of uncommitted tissue. Regeneration phenomena can be classified into those involving long-range interactions with extensive cell movements (morphallaxis) and short-range interactions with extensive growth (epimorphosis) (Morgan, 1901). Embryonic regulation, by contrast, involves no cell movement and no growth, but simply a readjustment of the fate map on a domain which is not yet determined.

Cytoplasmic determinants

A cytoplasmic determinant is a substance or substances, located in part of an egg or blastomere, which guarantees the assumption of a particular state of determination by the cells which inherit it during cleavage. In the past the existence of cytoplasmic determinants has often been deduced from quite inadequate evidence such as a predictable cell lineage (fate map), the occurrence of mosaic behaviour, or the presence in the egg of visible regional differences. All of these things are consistent with the presence of determinants but do not by any means prove their existence. The only really satisfactory proof is to transfer cytoplasm from one place to another by microinjection and show that cells inheriting the ectopic cytoplasm become structures normally formed by the egg region from which the cytoplasm came. This has been done in only one case: the pole plasm of *Drosophila*, which is discussed in Chapter 4. There are also a number of other experiments which will be described later in which the cytoplasm of eggs or early blastomeres is redistributed by centrifugation or compression with a resulting change in the pattern of the embryo. Some of these are fairly convincing and suggest at least that the cytoarchitecture of the egg may be important for later regional specifications. It should be noted that the significant regional differences seem sometimes to be established during oogenesis, sometimes after fertilisation and sometimes after one or more cleavages.

Granted that cytoplasmic determinants exist, it is rather unclear just what they represent at the biochemical level. At one extreme they might be regulatory molecules which are responsible for the derepression of specific genes at a certain time in development. At the other they might simply be a quantitative bias in the overall metabolism which is only the beginning of a complex chain of causation leading to differential regional

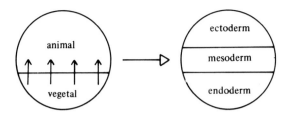

Fig. 2.6. An example of an instructive signal is provided by the induction of the mesoderm from the animal cap in response to a signal from the vegetal region in the early amphibian embryo.

specification. Biochemists often seem to assume that determinants are likely to consist of mRNA, although no reproducible regional differences have yet been found. Since there are usually many developmental decisions between the egg and the terminally differentiated cell types it seems unlikely that determinants would consist of mRNA coding for the proteins specific to the differentiated cells. If it were the case it would involve an extraordinary degree of 'molecular preformationism' in the egg and would not be compatible with the slightest degree of regulative behaviour. It seems more probable that determinants are responsible for the establishment of the first two or three distinctly specified regions and that subsequent regionalisation depends on interactions between these.

Induction

Regulation implies interactions, and among the regions which must be present for regulation to occur are the signalling centres or organisers. The signals which control the regionalisation of the early embryo are often called inductive signals, and the ability to respond to them is often called competence. To give a concrete example of this: there is some evidence that in the early amphibian embryo the mesoderm is formed from the animal cap tissue in response to some signal from the vegetal region (Fig. 2.6, and see also Chapter 3). The remainder of the animal cap becomes ectoderm, as does the whole animal cap in isolation. This interaction can occur between small pieces of the blastula cultured in combination, and so by using pieces taken from embryos of different stages it has been possible to show that the signal is present from morula to late blastula stages, while the competence of the animal cap tissue persists from the morula to the late gastrula stage. This type of interaction is sometimes called 'directive' or 'instructive' induction in so far as the responding tissue is faced with a choice (mesoderm or ectoderm?), and in normal development the interaction results in an increase in complexity of the embryo. The range of choices open to the competent tissue is a property of its state of determination. In this case once the mesoderm is formed it is competent to become somite, kidney, mesenchyme, etc., in response to further interactions, but it is no longer competent to become ectodermal derivatives such as neuroepithelium and epidermis. 'Competence' is thus a subset of 'potency', since it comprises all the outcomes achievable by a region of tissue in response to environments present in the embryo at the stage in question.

In this sense the 'instructive' character of inductive signals is limited by the competence of the responding tissue. This is seen clearly in experiments involving inductive interactions between tissues from different species, where the rule is that the structures formed by the responding tissue are always those of its own species rather than those of the inducer. For example, mouse dermis induces hairs from mouse epidermis but feathers from chick epidermis and scales from lizard epidermis (Sengel, 1976).

Some doubt has recently been cast on the reality of instructive signals and it is sometimes claimed that such interactions are simply 'permissive', i.e. the signal is necessary for the successful self-differentiation of the responding tissue but cannot influence the developmental pathway selected (Holtzer, 1978). Permissive interactions are certainly very important in late development and may also occur in early development. It is not conceivable, however, that the interactions which will be discussed in Chapters 3 to 6 of this book are permissive in character since they are clearly the foundation of the progressive regional subdivision and consequent increase in complexity of the body plan. If they are permissive then it means that some completely unknown process is responsible for generating the different types of cell and it also implies that normal development involves massive cell death and cellular migrations which have never been observed.

The mechanisms of inductive signals and the responses to them are hardly better understood now than when they were first discovered at the beginning of this century. If has often been supposed that such a signal must be a single chemical substance which diffuses from inductor to responding tissue. In the form of the gradient model presented in Chapter 9 this idea accounts for many of the facts, but is not necessarily the only model to do so. Very few substances are capable of diffusing rapidly across a sheet of cells if they have to cross the plasma membranes. This is because the membranes are hydrophobic and the soluble cytoplasm and extracellular matrix are hydrophilic. For this reason some attention has been concentrated in recent years on the so-called gap junctions which are widespread in early embryos as well as in adult tissue. These are channels running from one cell to another which are capable of passing low-molecular-weight substances by passive diffusion. They appear as distinctive hexagonal arrays of particles in freeze-fracture electron microscope pictures and can be detected experimentally by observation of the diffusion of a marker dye, or the passage of an electric current, from cell to cell (Feldman, Gilula & Pitts, 1978). If gap junctions are really the pathways for the inductive signals then it might be expected that the entire embryo would be connected ('coupled') at an early stage and that as regional subdivision proceeded it would become divided up so that each region was internally coupled but did not communicate with other regions. To some extent the results bear out this expectation although on the whole the subdivision into non-communicating zones seems to occur later than would be expected on the basis of what we know about the timing of inductive interactions.

It has also been suggested that inductive signals are mediated by contact between protein molecules in the plasma membranes of adjoining cells (McMahon, 1973), or by interactions between components of the extracellular matrix and the membrane of the responding cells (Grobstein, 1967). The evidence does not clearly support any one model and it is possible that all are right for different individual processes. In the latter cases, however, there is no single substance which diffuses across the responding tissue and apparent long-range effects are predicted to arise from a series of short-range effects. It is this sort of uncertainty about the basic character of the phenomenon which has prevented for so long any molecular-level understanding of induction, because it is not possible to design an assay for any component of the system without making a good number of unjustified assumptions about how the system works as a whole.

The embryonic field

Like 'polarity' the term 'field' has two clearly distinguishable usages in modern embryology which we might term the physical and the agricultural forms. In physics a field is a variable, scalar or vector, which varies in space. By analogy with this we could refer to an embryonic field and mean the equations which define the values in different parts of the embryo of the key biochemical variables which control regional specification. We shall come across some hypothetical equations of this type in later chapters. Usually, however, the agricultural analogy is what people have in mind when they refer to a field; in other words it is the area of tissue within which a certain process, such as an inductive interaction, occurs. So for example the 'limb field' is that part of the mesodermal mantle in which interactions occur around the time of gastrulation which result in the formation of a determined limb rudiment. It follows that it is not proper to refer to a limb field at a stage before the relevant interactions commence. There is no limb field in the fertilised egg, simply a prospective region for the limbs.

According to this definition, an embryonic field is the same thing as a 'polyclone' in the development of *Drosophila* (Crick & Lawrence, 1975) or an 'equivalence group' in the development of the nematode *Caenorhabditis* (Kimble, 1981). Those who work on these organisms evidently feel that the classical term 'field' is too vague to be of use, and in view of the commonly encountered ambiguity this is understandable.

Part II

The data base

3

The amphibian embryo:
a hierarchy of developmental decisions

Due to a number of technical advantages which they offer, amphibian embryos have been a favourite material for the experimentalist for over 100 years. The eggs are large, usually 1–2 mm in diameter, and therefore suitable for micro-operative procedures. They are laid in jelly coats simultaneously with fertilisation and complete their entire development to the hatching larva within these coats, so remaining accessible to experimentation at all stages. In addition, as we shall see below, fragments of early embryos are able to continue development in isolation if incubated in simple isotonic or hypotonic salt solutions. This is possible because every cell in the embryo contains a supply of yolk platelets which serve as its nutrient supply until the larval blood circulation becomes established.

Many different species have been used for experimental work. The 'classical' work of the first half of the twentieth century used mainly embryos of the European newts *Triturus* or frogs *Rana*, which were collected in the field in each breeding season. More recently there has been a move towards animals which can easily be bred in the laboratory (Table 3.1). The most popular are probably *Xenopus laevis*, the African clawed toad, *Rana pipiens*, the American leopard frog, and *Ambystoma mexicanum*, the axolotl. Taxonomically the frogs and toads are called Anura and the newts and salamanders are called Urodela (Greek *oura* = tail). Most amphibians hatch as larvae which feed and grow for a few months before metamorphosing to the adult form, although the axolotl retains a larval morphology throughout its life. Amphibians have of course been extensively used for research into the mechanisms of regeneration and metamorphosis, but we are here concerned only with their early embryology. Most of the principal results which will be discussed below have been obtained on more than one species, but the species used should be taken into account when reading the literature, particularly with regard to the tempo of development; *Xenopus*, for example, develops to hatching in a few days while most other species take 2–3 weeks.

Table 3.1. *Amphibians used in embryological research*

Species	Common name	Origin	Country of use
Frogs			
Rana pipiens	American leopard frog	USA	USA
R. temporaria (*=fusca*)	Common European frog	Europe	Europe
Xenopus laevis	African clawed toad	Africa	World-wide
Salamanders			
Triturus vulgaris (*=taeniatus*)	Common newt	Europe	Europe
T. alpestris	Alpine newt	Europe	Europe
T. cristatus	Italian newt	Europe	Europe
Ambystoma punctatum (*=A. maculatum*)	Spotted salamander	USA	USA
Ambystoma mexicanum (*=Siredon mex.*)	Axolotl	Mexico	World-wide
Pleurodeles waltl	Ribbed newt	Algeria	France
Cynops (*=Triturus*) *pyrrhogaster*	Japanese newt	Japan	Japan

Normal development

When laid the amphibian egg has a dark pigmented animal hemisphere and a light-coloured vegetal hemisphere. It lies within a transparent vitelline membrane inside the jelly coat. After fertilisation the membrane lifts from the egg surface and the egg rotates under the influence of gravity so that the less dense animal hemisphere lies uppermost. As in other types of embryo the two polar bodies appear at the animal pole and contain the unused chromosome sets from the first and second meiotic divisions. The first polar body is produced after maturation and the second after fertilisation. The axes of the egg are shown in Fig. 3.1 and stages of early development in Fig. 3.2.

Anuran eggs are usually fertilised by a single sperm, urodele eggs by more than one, although only one sperm pronucleus actually fuses with that of the egg. Shortly after fertilisation the 'grey crescent' appears, in monospermic eggs opposite the site of sperm entry. The grey crescent marks the prospective dorsal side although it is not visible in all species and indeed its visibility varies considerably between batches of eggs. The first cleavage is vertical and usually bisects the grey crescent, separating the egg into right and left halves. The second cleavage is at right angles to this and separates dorsal from ventral halves. The third cleavage is equatorial, separating animal from vegetal halves. As in other embryos,

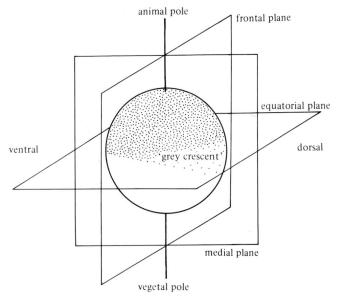

Fig. 3.1. Axes of the amphibian egg after fertilisation.

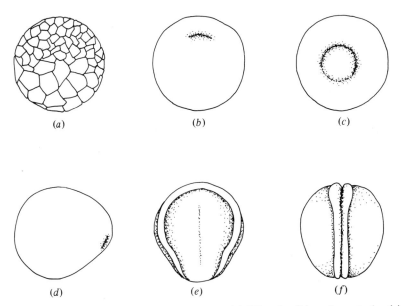

Fig. 3.2. Stages of amphibian development. (*a*) Blastula, (*b*) early gastrula, (*c*) late gastrula, (*d*) slit blastopore, (*e*) early neurula, (*f*) closed neural tube. (*a*) is a view from the side. (*b*) and (*c*) are views from the vegetal pole, (*d*) is a view from the side, and (*e*) and (*f*) are dorsal views.

the large cells resulting from early cleavage divisions are called blasto-meres. As cleavage takes place a cavity called the blastocoel forms in the centre of the animal hemisphere and the embryo is referred to as a blastula. Cleavage continues rapidly for about 12 to 15 divisions, synchrony being lost and the division rate slowing towards the end of this phase.

The next phase of development is one of extensive morphogenetic movements called gastrulation (Fig. 3.3). A depression appears in the dorsovegetal quadrant, becomes elongated laterally and finally becomes a complete circle. This is the blastopore, which being circular can be said to have dorsal, lateral and ventral lips. In urodeles the blastopore represents a locus of invagination of a belt of tissue stretching around the equator of the blastula: the marginal zone. This is the prospective mesoderm, and the animal region which does not invaginate is the prospective ectoderm. Dorsally the invaginated material is double-layered, as though a finger has been pushed into a balloon, and the new space which appears between these two layers is called the archenteron. The dorsal invagination proceeds until the leading edge of tissue lies in the vicinity of the animal pole, with the archenteron roof being closely apposed to the overlying ectoderm. In the lateral and ventral parts of the blastopore the inward movement is of a solid mass of cells, the ventral ingression being slight and the lateral ingression joining the dorsal invagination near the animal pole. Simultaneously with the invagination movements, the animal cap tissue spreads vegetally so that the blastopore lip becomes a smaller and smaller ring around the vegetal pole and eventually narrows to a small slit.

So by the end of gastrulation the former animal cap ectoderm has covered the whole external surface of the embryo, the yolky vegetal tissues have become a mass of endoderm in the interior, and the former marginal zone has become an intermediate layer of mesoderm extending anteriorly from the slit-shaped blastopore as a cylinder whose open end continues its movements to cover the antero-ventral region in later stages.

The gastrulation movements in the Anura differ somewhat in that the mesoderm delaminates from the animal hemisphere in the blastula and then turns inside out and elongates dorsally to form the definitive mesodermal mantle. The archenteron is formed by invagination but its tip reaches well beyond the former animal pole. The end result, however, is much the same as in urodeles.

By the end of gastrulation the archenteron has become the principal cavity at the expense of the blastocoel, and the embryo has rotated so that the dorsal side is uppermost. It now has a true anteroposterior (or craniocaudal) axis which runs from the leading edge of the mesoderm to the blastopore. This corresponds roughly but somewhat fortuitously with the original animal–vegetal axis. The three germ layers have now reached

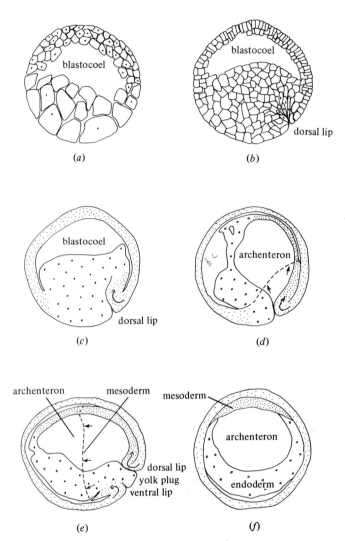

Fig. 3.3. The process of gastrulation in a urodele amphibian. (*a*) to (*e*) are shown in median section and (*f*) in transverse section. The dashed line in (*d*) and (*e*) represents the anterior limits of lateral mesodermal ingression.

their definitive positions: ectoderm outside, endoderm inside and mesoderm in between.

The next phase of development is called the neurula (Fig. 3.4), in which the ectoderm on the dorsal side becomes the central nervous system. The neural plate becomes visible as a keyhole-shaped region delimited by raised neural folds and covering much of the dorsal surface of the embryo. Quite rapidly the folds rise and move together to form the neural tube

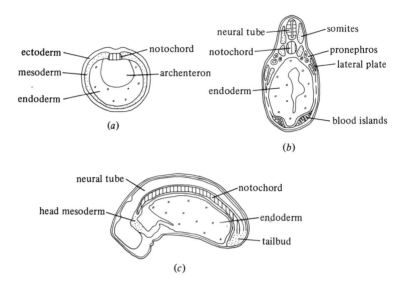

Fig. 3.4. Later stages of amphibian development. (*a*) Transverse section through a slit blastopore stage. (*b*) Transverse section through a urodele tailbud stage. (*c*) Median section through a urodele tailbud stage.

which becomes covered by the ectoderm from beyond the folds, now known as epidermis. Tissue from the folds which comes to lie dorsal to the neural tube is the neural crest. In the mesodermal layer the posterior part of the dorsal midline segregates as a distinct notochord, and the mesoderm on either side begins to become segmented in anteroposterior sequence to form paired somites. The endoderm sends folds up beneath the archenteron roof which meet beneath the notochord so that the entire archenteron cavity, the prospective gut, is now lined with endoderm. There is throughout this period a general dorsal concrescence of the mesoderm and anteroposterior elongation of the entire embryo.

The end of neurulation is a convenient place for us to stop since by this stage the rudiments for all the major structures of the body are in their definitive positions. The central nervous system is formed from the neuroepithelium of the neural tube. The mesoderm of the trunk region develops into several structures in dorsoventral sequence: notochord, myotomes, pronephros and mesonephros, and ventral blood islands. Anteriorly the prechordal mesoderm forms part of the jaw muscles and branchial arches. The heart, limbs and future haemopoietic system also exist as rudiments in the mesodermal mantle. The endoderm becomes the epithelial components of the pharynx, stomach, liver, lungs (if any), intestine and rectum, and the tail arises from the tailbud, which cannot properly be considered exclusively ectodermal or mesodermal since it produces neural tube, somites and mesenchyme.

There are several important cellular migrations which occur in later development. Cells from the neural crest become much of the head, autonomic nervous system and melanophores of skin and gut. The primordial germ cells migrate dorsally from the germinal ridges to the gonads, the haemopoietic cells move from the mesonephric region to the liver, and somitic myoblasts supply the striated muscle of ventral body wall and limbs. However, the basic body plan is not disturbed by these migrations and presumably provides the scaffolding on which they take place.

Despite the fact that the basic body plan is determined, it should be noted that there is very little visible histological differentiation by the end of neurulation. All the cells of an amphibian embryo at this stage are still packed with yolk granules. The tissue layers can be distinguished to some extent by the shapes of their cells but mainly by their mutual relationship in the normal pattern. This is why we refer to organ rudiments rather than organs. It should also be noted that although an amphibian embryo does become a little bigger over the period covered, this is entirely due to the uptake of water. The embryo has no supply of nutrient from an extra-embryonic yolk mass or a placenta and therefore loses dry weight by metabolic activity throughout development. All cell divisions until quite a late stage are cleavage divisions in which daughter cells are half the size of the mother cell. This is true of all holoblastic eggs which develop independently of the mother, but is often forgotten by mammalian or chick embryologists whose organisms undergo growth and development simultaneously.

Fate maps

Amphibians possess a fate map which is continuous from the fertilised egg onwards, showing that there is no stage of random cell mixing. Most of the experimental work involved in mapping the prospective regions and in tracing the morphogenetic movements has been carried out by the technique of vital staining. In this method a small piece of agar impregnated with the stain is pressed against a particular region on the surface of the embryo until the cells have become deeply stained. The embryo is then allowed to develop to a later stage, is fixed and dissected and the location of the stained cells noted. Internal regions can be stained by using agar spikes which are inserted to a certain depth and stain all the cells around them. The account given above of the gastrulation movements is based on the classic work of Vogt (1929) as added to by Pasteels (1942), Keller (1975, 1976) and other authors.

In Fig. 3.5 are shown the trajectories of a number of marks applied to the dorsal meridian. Those near the blastopore move in sequentially through the dorsal lip and eventually end up disposed in a continuous ring

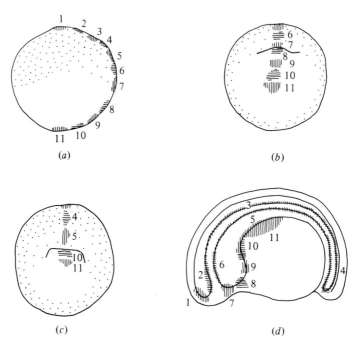

Fig. 3.5. Movement of vital stain marks applied to the external surface of early urodele embryos. (a) The marks immediately after application to the blastula. (b) and (c) The invagination of the marks through the dorsal lip. (d) The positions of the marks in the tailbud stage. (After Vogt, 1929.)

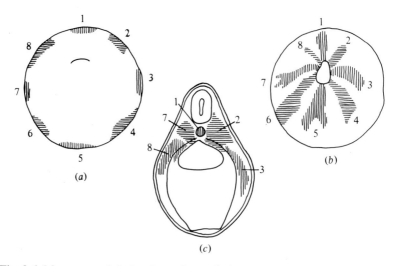

Fig. 3.6. Movement of vital stain marks applied to the marginal zone of an anuran gastrula. The marks are originally equally spaced around the equator. They invaginate through the blastopore and end up as longitudinal stripes concentrated towards the dorsal side of the mesodermal mantle. (After Vogt, 1929.)

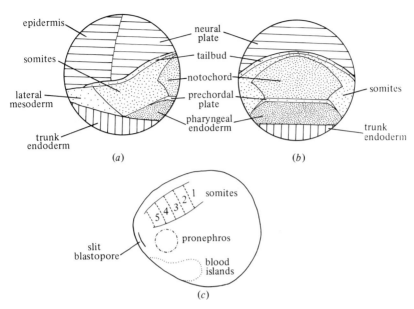

Fig. 3.7. Fate maps of the early amphibian embryo compiled from vital staining data. (*a*) and (*b*) The urodele blastula from the lateral and the dorsal aspects with the prospective regions in the surface layer indicated. (After Pasteels, 1942.) (*c*) Prospective regions in the mesoderm of a urodele slit blastopore stage. (After Yamada, 1937.)

around the long axis of the archenteron. The marks near the animal pole never reach the blastopore and so end up inside the neural tube. In Fig. 3.6 a similar experiment is depicted in which equatorial marks are used. These become greatly elongated as the marginal zone enters through the ring-shaped blastopore but dorsoventral continuity is maintained in the mesodermal mantle as shown in the transverse section of the tailbud stage (the ventral marks are present only posterior to the section shown). Fig. 3.7 shows a fate map of a urodele gastrula compiled from many such marking experiments. All the prospective regions are shown on the surface but it is probable that the mesodermal regions extend to the interior. In the Anura the prospective lateral and ventral mesoderm is entirely internal at this stage. The boundaries between the prospective regions are shown as sharp lines but they are probably somewhat fuzzy on account of local cell mixing. Fig. 3.7(*c*) shows some prospective regions in the mesodermal mantle of the slit blastopore stage and it will be noted that much of the posterior part of the body is not represented, being as yet still not produced from the tailbud.

Vital staining is only applicable to clumps of cells and because the stain fades and becomes diluted it is not possible to say that every cell follows a strictly predetermined trajectory, simply that small regions do so.

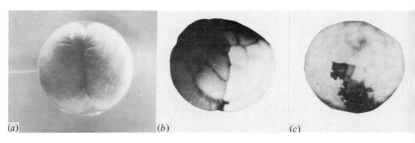

Fig. 3.8. The horseradish peroxidase technique for fate mapping. (*a*) A *Xenopus* embryo at the 2-cell stage with a microinjection needle inserted into one blasto-mere. (*b*) A morula stained for horseradish peroxidase after injection of one blastomere at the 2-cell stage. The left half of the embryo is stained. (*c*) Vegetal view of a blastula showing a clone derived from one blastomere injected at the 32-cell stage. (Author's photographs.)

Recently the method of intracellular injection of horseradish peroxidase has been applied to the *Xenopus* embryo by Hirose & Jacobson (1979) and Jacobson & Hirose (1981). Amphibian embryos are well suited to a passive marking technique of this sort since the label is not diluted by growth. The enzyme is reasonably stable after injection and unlike the vital dyes is not passed to neighbouring cells. It can be detected in single cells at a later stage by the very sensitive histochemical methods for peroxidase (Fig. 3.8). This has yielded results for the central nervous system which are of higher resolution than the fate maps elucidated by vital staining, although not incompatible with them. The method has confirmed suspicions that there is some cellular mixing at the edge of each clone, this being particularly noticeable in the ventral epidermis of the neurula in which there is extensive mixing of cells across the midline (author's unpublished results).

If it is shown that a clone remains coherent and ends up entirely within a particular structure at a later stage of development this does not of course mean that the cell was committed to form that structure at the time it was labelled, since in embryos which have a fate map every structure must have some predictable prospective region at an earlier stage quite irrespective of any developmental decisions which may be made. How-ever, as was discussed in Chapter 2, the converse is valid: if a clone crosses the boundary between two structures A and B, this means that the cell was not irreversibly committed to become either A or B at the time of labelling. So if there is a reasonable amount of cellular mixing locally some information about states of determination can be obtained by clonal analysis of cells which lie near prospective boundaries. At the time of writing no such data have yet been published, although we may expect it to appear soon since the problem is an important one and the techniques are now available.

Regionalisation within the egg

It is generally agreed that the unfertilised egg is radially symmetrical, having no prospective dorsal or ventral side. Along the animal–vegetal axis (egg axis) there are four inhomogeneities which become established by unknown mechanisms during oogenesis. These are a deep pigmentation of the superficial layer of the animal hemisphere; a difference in size of yolk granules (small in the animal and large in the vegetal hemisphere); the presence of more non-yolk cytoplasm in the animal hemisphere; and the presence of the nucleus, or germinal vesicle, in the animal hemisphere. On maturation the germinal vesicle ruptures and one diploid chromosome set is expelled at the animal pole as the first polar body. After fertilisation the second meiotic division is completed and a second chromosome set expelled as the second polar body. Whether any of these visible inhomogeneities are causally related to the later regionalisation of the egg axis and whether there are further unknown inhomogeneities is not known.

The establishment of a prospective dorsoventral axis occurs shortly after fertilisation (Nieuwkoop, 1977; Brachet, 1977). In some species the so-called grey crescent appears on the prospective dorsal side, perhaps as a result of local retraction of the pigmented cortex. It was first shown in the nineteenth century and confirmed by Ancel & Vintemberger in 1949 that in anurans, which are monospermic, the grey crescent normally appears opposite the point of sperm entry, which is marked by a small pigmented spot. However, a grey crescent also appears in eggs parthenogenetically activated by pricking with a fine needle, and is then not necessarily opposite the pricked point. Also many urodeles are polyspermic and sperm pits may appear anywhere on the embryo, with the grey crescent and future dorsal side not being related to them in any particular way. Moreover the dorsoventral polarisation induced by the sperm may subsequently be overridden by other stimuli, as will be described shortly. The establishment of the dorsoventral axis is therefore thought to be an example of a symmetry-breaking process in which an unstable homogeneous steady state spontaneously evolves into a stable inhomogeneous steady state. The symmetry of the final state is determined by any of a variety of possible minor perturbations, such as in this case the disturbance occasioned by the sperm. Symmetry-breaking processes are further discussed in Chapter 8.

Although in normal development of the Anura the grey crescent and later the blastopore appear opposite the point of sperm entry, it is possible to alter the prospective dorsoventral axis after fertilisation by rotating the egg through 90 degrees (Fig. 3.9). This can readily be done by immersing the eggs in a solution of Ficoll, a high-molecular-weight carbohydrate which cannot penetrate the vitelline membrane. Water is

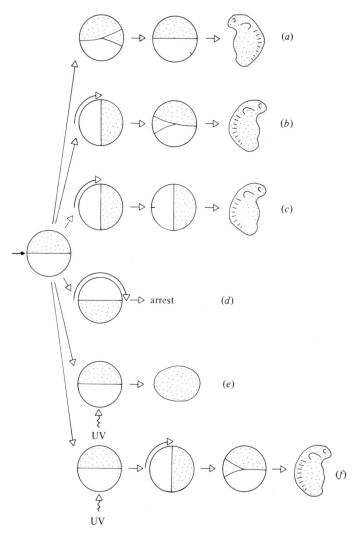

Fig. 3.9. Experiments on determination of dorsoventral polarity in the early amphibian embryo. (*a*) Normal development in which the dorsal side develops opposite the point of sperm entry. (*b*) Inversion of the axis by a 90 degree rotation after fertilisation. (*c*) Alteration of the animal–vegetal axis of the embryo by sustained rotation through 90 degrees. (*d*) Arrest of development following sustained rotation through 180 degrees. (*e*) Symmetrisation of the embryo by irradiation of the vegetal pole with ultraviolet (UV) light shortly after fertilisation.

(*f*) 'Rescue' of UV-irradiated embryo by subsequent 90 degree rotation.

removed from the space below the membrane so that the egg can no longer rotate freely within its jelly coat. Then if the egg is rotated through 90 degrees, or any other angle, it will stay put and internal cytoplasmic rearrangements take place in response to the gravitational field. If a 90-degree rotation is carried out up to 75 minutes post-fertilisation in *Xenopus*, which normally cleaves at 90 minutes post-fertilisation, then the dorsal side appears on the face which was uppermost (Kirschner *et al.*, 1980; Gerhart *et al.*, 1981). If this conflicts with the side opposite the point of sperm entry then the rotation overrides the sperm. A similar repolarisation can be achieved by a slight temperature gradient (Glade, Burrill & Falk, 1967).

The normal visible events involved in dorsoventral polarisation are: (1) formation of the grey crescent, perhaps by displacement of the pigmented cortex; (2) formation of a region of clear (yolk-free) cytoplasm in the centre of the animal hemisphere and its movement to the prospective dorsal side; (3) formation of a tongue of yolk adhering to the cortex on the prospective dorsal side, the 'vitelline wall'. After rotation, the principal vegetal yolk mass slips downwards and it is probable that displacements similar to 2 and 3 above would be initiated. Since rotation will repolarise up to but not beyond formation of the grey crescent it is probable that the gravitational movements are sufficient to break the symmetry of the newly fertilised egg, and the effects of the sperm are overridden because its perturbation is even weaker. But the definitive dorsoventral pattern of cytoplasmic regionalisation is brought about by cytoplasmic movements inherent to the structure of the egg, not by the effect of gravity. In other words, the dynamical properties of the egg are the 'formal cause' and the sperm or gravity merely the 'efficient cause' (see Chapter 8).

A stimulus which is known to disrupt these movements is ultraviolet (UV) irradiation. If fertilised eggs are irradiated from the vegetal side the resulting embryos lose dorsal structures. This effect is dose-dependent, and the most effective wavelength is 280 nm, which suggests a protein target (Malacinski, Benford & Chung, 1975; Woo Youn & Malacinski, 1980). Since the radiation only penetrates a short distance it has relatively little effect on cellular viability. It is now known that 90-degree rotation after a UV dose can completely 'rescue' the embryo and allow it to form a normal pattern (Fig. 3.9, and Scharf & Gerhart 1980). So it is thought that the UV effect involves destruction of a component required for the dorsoventral polarisation rather than destruction of a cytoplasmic determinant for the dorsal region.

Cleavage

In anurans the first cleavage plane is normally medial and bisects the grey crescent; the second is frontal and the third equatorial. A deterministic sequence of cleavages inevitably means that particular regions of zygote cytoplasm will be included in particular blastomeres and inhomogeneities will become differences between blastomeres. In urodeles the first cleavage may not be medial, and so it is possible for the grey crescent to lie entirely within one of the two blastomeres. The second cleavage is always also vertical, and orthogonal to the first, and the third is equatorial. If embryos are lightly compressed under a glass plate it is possible to cause the third and fourth cleavages to occur in vertical orientation in preference to an equatorial one. However, such embryos will develop normally once the glass plate is removed (see Morgan, 1927). It seems, therefore, that the positioning of plasma membrane relative to cytoplasmic regions at this stage is not critical for the operation of subsequent developmental decisions.

There is some evidence, though, that disturbance of the normal relative arrangement of regions in the zygote cytoplasm can affect viability and/or cause pattern abnormalities. If eggs are maintained in a 90-degree rotated position throughout early development then the dorsal blastopore lip appears near the original vegetal pole and the parts of the embryo are drastically misplaced relative to the cortical pigmentation, although as discussed above the internal cytoplasmic regions may be similar to the normal. *Rana* embryos arrest after gastrulation although axolotl embryos may still go on to develop normally (Malacinski & Chung, 1981). If eggs are maintained upside down (180-degree rotation) then both axolotls and *Rana* arrest at an early stage (Fig. 3.9).

Artifical inversion of eggs has also been known for some time to result in the production of double embryos with two blastopores (Pasteels, 1938), and it has recently been shown that the same can occur even at the 2-cell stage if embryos are centrifuged at 30–50 g towards the ventral side (Gerhart *et al.*, 1981). The phenomenon is probably associated with a breaking up of the yolk mass. The resulting embryos resemble the partial twins that may be produced by a variety of other methods such as partial constriction with a hair loop at the 2-cell stage, or organiser grafts at the early gastrula stage (see later). They are characterised by the presence of two separate sets of dorsal axial structures over at least part of the anteroposterior axis with a plane of mirror symmetry in between them. The significance of the structure will be discussed later but it does suggest that at least the reference points for the future pattern-forming signals are established at this stage and bear some correspondence to the arrangement of cytoplasmic regions after dorsoventral polarisation in the zygote.

Further evidence for this comes from the classical constriction experiments of Spemann (reviewed by Spemann, 1938). The first two blastomeres of an amphibian embryo are usually tightly apposed to one another. But Spemann managed to separate them by tightening a hair loop around the embryo in the plane of the first cleavage. He found, using

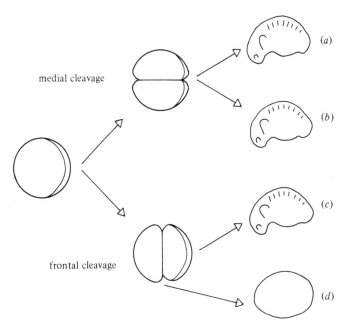

Fig. 3.10. Production of twins by separation of the first two blastomeres. If cleavage occurs in the medial plane both blastomeres produce an embryo (*a*, *b*). If cleavage occurs in the frontal plane the ventral blastomere can produce a symmetrical 'belly piece' (*c*, *d*).

Triturus vulgaris which does not have a readily visible grey crescent, that often each blastomere could develop into a complete, normally proportioned, but half-sized embryo. Sometimes, however, one blastomere formed a complete embryo but the other formed a 'belly piece' similar to the structures formed by eggs after UV irradiation. He concluded that the belly pieces arose from the ventral blastomeres of frontally cleaving embryos and therefore that some dorsal cytoplasm was necessary for the establishment of a complete pattern (Fig. 3.10).

Another aspect of regionalisation in the fertilised egg is the possible existence of a germinal determinant in the Anura (Smith, 1966; Whittington & Dixon, 1975; Nieuwkoop & Sutasurya, 1977). Anuran eggs contain a region of special cytoplasm in the vegetal cortical region which is later found in the primordial germ cells. In *Rana pipiens* irradiation of the vegetal hemisphere with low doses of UV light can produce larvae lacking germ cells but otherwise apparently normal. The action spectrum for this effect shows a peak at 260 nm, suggesting a nucleic acid target. Eggs can be 'rescued' by subsequent injection of vegetal but not animal cytoplasm, although this is not quite sufficient proof for the existence of a determinant because it is not known whether the germ cells in the rescued hosts actually inherit the donor cytoplasm. It should also be noted that the

UV doses used are sufficient to cause some somatic pattern defects in other species as discussed above. However, the evidence certainly suggests that some early event in germ cell determination depends on something in the vegetal region. Germ cell determination in urodeles seems to occur somewhat later and as a result of inductive interactions (Nieuwkoop & Sutasurya, 1977).

States of commitment in the blastula

The nuclei of blastula cells do not appear to be restricted in potency since one nucleus can support the development of an entire embryo. This was first shown by Spemann in 1928 (reviewed by Spemann, 1938). A hair loop was used to constrict a fertilised egg of *Triturus* into a dumb-bell shape in which one half contained the zygote nucleus and the other half did not. Cleavage takes place only in the nucleated half but if the loop is not too tight a nucleus may pass through the constriction up to the 64-cell stage and initiate development in the previously enucleated half. These halves developed into complete embryos despite the fact that their nucleus had already undergone several rounds of DNA replication and mitosis. A similar result was obtained by Briggs & King (1952) who injected blastula nuclei into enucleated host zygotes. This procedure has subsequently been used to 'clone' amphibians, since a number of embryos can be grown using nuclei from a single blastula (Gurdon, 1974; McKinnell, 1978).

It is unfortunate, and to some extent surprising, that we have no information about the states of specification or determination of whole single cells in early amphibian development, but modern techniques of cell marking and micromanipulation should enable such data to be available soon.

There is, however, a great deal of evidence about the states of specification and determination of small regions of tissue. Specification is assayed by explantation of small regions followed by culture in buffered salt solutions, a method introduced by Holtfreter (1938a, b). Amphibian embryos have the great advantage from the experimenter's point of view that every cell contains some yolk platelets and these can serve as its nutrient source until after the larva hatches. This means that various cell types will differentiate if explants are maintained in buffered salt solution for a sufficient length of time. The cells of the explant continue to cleave and so increase in number, but since there is no external nutrient supply they do not increase in dry weight. Actually the advantages of this method are less than they appear at first sight, for three reasons:

Firstly, explants usually give rise to more than one cell type. This is because, as mentioned in Chapter 1, histologically recognisable cell types are formed as a result of late local interactions rather than the interactions which partition the early embryo into regions. Fortunately in amphibian embryos some cell types develop only from certain subdivisions of the body and so in the case of an explant it is often possible to assign it a state of specification on the basis of the cell types which are produced. Secondly, the assumption that the buffered salt solution is a 'neutral' medium may perhaps be justified, but it is not so clear that it is a 'simple' medium. Explants do not usually differentiate well unless they become surrounded by epidermis, either produced from the explant itself or supplied by the experimenter. So the microenvironment within the epidermal vesicle is an unknown quantity. Thirdly, single dissociated cells do not differentiate as well as clumps of tissue, even when the clumps are not surrounded by epidermis. Only at neurula and later stages will single cells differentiate reliably (Elsdale & Jones, 1963). So presumably something produced by the cells themselves is necessary for their proper development.

It is important to understand the drawbacks of this method not because it is without value, but because it has been of such decisive importance in understanding early amphibian development and it is largely because of it that the dynamics of the early amphibian embryo are so much better understood than those of mammals or birds.

From morulae up to about the 64-cell stage explants from the animal hemisphere will develop into balls of ciliated epidermis while explants from the vegetal hemisphere fail to differentiate. From this stage onward explants from the equatorial region (the marginal zone) will develop into mesodermal tissues, indicating that a mesodermal region has been specified (Nakamura & Matsuzawa, 1967; Nakamura & Takasaki, 1970). From the earliest stage at which this is possible there is a difference between dorsal and ventral marginal zone. Explants from dorsal marginal zone form notochord, muscle and neuroepithelium, and those from ventral marginal zone tend to form epidermis, mesenchyme and erythrocytes (Fig. 3.11). These are the tissues that arise from the dorsal and ventral tissues in the course of normal development.

Experiments carried out by P. D. Nieuwkoop and his coworkers (reviewed by Nieuwkoop, 1973) indicate that this mesodermal rudiment is brought into existence as a result of an interaction along the egg axis (Fig. 3.12). Axolotl blastulae were divided into four zones along the egg axis, with I and II drawn from the animal cap, III being the marginal zone, and IV the vegetal yolky region. When cultured alone I and II tended to form epidermal bags and IV showed poor differentiation. But III develops into a deformed but not unreasonable complete embryo. When the regions were recombined and cultured it was found that the combination I + II + IV also yielded a reasonable version of a complete embryo (Fig. 3.12). When the orientation of IV was altered relative to I + II it could be shown that the dorsal structures developed on the side which was

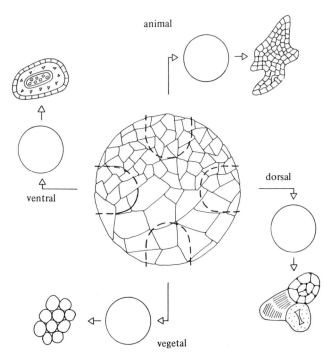

Fig. 3.11. Self-differentiation of isolates from different regions of the amphibian blastula. Isolates from the animal cap produce only epidermis, isolates from the vegetal region fail to differentiate, isolates from the dorsal marginal zone produce notochord, muscle and neuroepithelium, while isolates from the ventral marginal zone produce epidermis, mesenchyme and erythrocytes.

dorsal in IV, irrespective of the orientation of I and II. This is despite the fact that the grey crescent itself is mainly confined to II and III. When the I + II + IV recombination was assembled from marked components it was established that the entire mesoderm and the pharyngeal endoderm were derived from the I and II section. So, it seems that even after the mesodermal rudiment has been specified, it can still be reconstituted from its surroundings, or if isolated it can still reconstitute the remainder of the embryo.

There is some terminological disagreement about the process. Nieuwkoop calls it 'induction of the mesoderm by the endoderm' while Nakamura refers to 'regulation of a gradient'. The data, however, are substantially agreed upon.

The interaction is mimicked by two agents which are active on isolated blastula or gastrula ectoderm: lithium ion and the 'mesoderm inducing factor'. This effect of lithium is only one of several effects it has on early amphibian development (see below) and indeed of effects on early

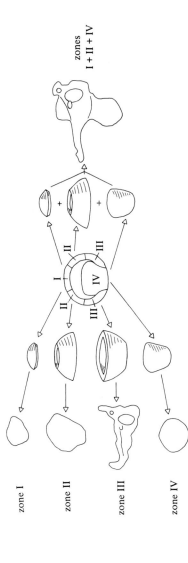

zones
I + II + IV

zone I

zone II

zone III

zone IV

Fig. 3.12. Development of annular fragments from the axolotl blastula. Only zone III can produce a reasonably complete embryo in isolation, but a similar embryo can be produced from zones I, II and IV in combination.

embryos in general. Masui (1961) reported that isolates of *Triturus* gastrula ectoderm would develop into mesodermal tissues after treatment with lithium chloride. Nieuwkoop (1971) is of the opinion that whole embryos or ectoderm–endoderm combinations can be vegetalised by lithium but isolated ectoderm cannot.

The existence of the 'mesoderm inducing factor', here to be referred to as M factor, was first inferred by Chuang (1939) when he found that adult kidney contained a heat-labile mesoderm inducing activity as well as a heat-stable neural inducing activity (see below). Since then much effort has gone into the isolation of the active principle (Tiedemann, 1976). The current M factor is purified protein of molecular weight 30 000 isolated from late (11–13-day) chick embryos. Blastula or early gastrula ectoderm will develop into mesodermal tissues after treatment and it is apparently also possible to determine which mesodermal tissue predominates by varying the time of treatment, or by combination with less pure preparations (Asahi *et al.*, 1979). Notochord and muscle are indicative of a dorsal mesoderm state of specification and mesenchyme and blood of a ventral mesoderm state.

So it seems that there are two agents which will achieve *in vitro* what the normal egg axis interaction does *in vivo*: namely to transform ectoderm into mesoderm or even endoderm. It is thus somewhat disappointing that the biochemical investigations have not proceeded further than they have. The biochemical action of lithium is not known, although in adult brain tissue it is supposed to accelerate metabolism of noradrenaline and serotonin. Serotonin is present in the early amphibian embryo in small amounts (Baker, 1965) but shows no significant regional distribution and is unaffected by treatment with lithium (D. Forman & J. M. W. Slack, unpublished results).

The M factor is obtained from late chick embryos or other heterologous sources. It has been claimed that a similar activity can be extracted from early amphibian embryos but it has not been purified, and it must remain an open question whether it is really a similar protein or not. It seems unlikely that a 30 000 molecular weight protein would itself be a signal substance in an embryo as large as an amphibian where we might expect the time required for diffusion to be limiting. It is known that the amphibian blastula is coupled electrically, suggesting that gap junctions connect all cells (Slack & Warner, 1973), so it is tempting to look for the signal substances among the small molecules which can pass through the junctions, and it is possible that the M factor might affect the production or the removal of such substances.

In any case it is clear that during its development from early cleavage to late blastula the amphibian embryo acquires several distinctly specified regions. To begin with there are probably only two: animal and vegetal regions, perhaps formed simply by passive partition of the egg cytoplasm.

By the late blastula there is an ectoderm, mesoderm and endoderm, and the mesoderm is divided into at least two zones in the dorsoventral axis. But we also know of various regulative properties possessed by the system of territories in the blastula. First we will consider the effect of changing the size of the embryo and then enquire which particular regions of the blastula are necessary for the reconstitution of the total pattern.

Regulation

The most straightforward way of changing the size of an embryo is by removing parts of the fertilised egg. Kobayakawa & Kubota (1981) report experiments in which fertilised eggs are slowly bisected by a glass rod into a nucleated and enucleated half. This procedure can be repeated to produce embryos having only a quarter of the normal volume. The enucleated fragments do not develop, but the nucleated ones do develop and produce normally proportioned embryos. Similar conclusions can be drawn from the classical constriction experiment involving separation of the first two blastomeres, at least one and usually both forming normally proportioned embryos. Cooke (1975, 1981) has made quantitative studies of somite number and of the number of cells in the subdivisions of the mesoderm and has shown that reduction of blastula volume by about 50% does not affect the proportions significantly.

It is also possible to create double-size embryos by the aggregation of two 2-cell stages (Mangold & Seidel, 1927). This procedure frequently yields embryos with multiple axes, but a single embryo can be formed when one embryo with medial cleavage is fused with one with frontal cleavage in such a way as to align the dorsoventral axes. Such embryos are twice the normal size and are at least approximately normally proportioned.

Taking these experiments together we see that it is possible to vary the volume of an amphibian embryo by a factor of 8, and hence vary each linear dimension by a factor of the cube root of 8, which is 2. Over this range the size of each part of the body is scaled to the size of the whole and each part is composed of a different number of normal-sized cells from the usual. This conclusion is not affected by whether the volume change is made before the territories are established or while they are being established.

It is also possible to vary the cell size but keep the total embryo volume constant. Application of pressure of about 6000 p.s.i. around the time that the second polar body is expelled can cause its resorption and, often, its incorporation into the zygote nucleus along with male and female pronuclei. Such embryos then have three chromosome sets and are called triploid. Triploid cells are 50% larger than diploid ones but are otherwise normal and able to carry out mitosis. Pressure shocks, and also hot and cold shocks, can on occasion give rise to embryos of other ploidy ranging from haploid to heptaploid (Fankhauser, 1945). These animals have been studied exhaustively by Fankhauser and Humphrey and it seems clear that although the embryos are of normal size, each structure within them is composed of a different number of cells from usual. In haploid embryos there are twice as many cells as usual and in polyploids there are fewer than usual in inverse proportion to the increase in cell volume.

So the size of parts does not depend on the size of individual cells, nor does it depend on some absolute dimension specified in the developmental program; rather it depends on the size of the whole embryo. This suggests that the specification of territories depends on signals which extend across the whole extent of the embryo and can scale each territory to the amount of tissue available.

The second important aspect of regulation is the regional variation in regulative ability. In the early stage it seems probable that it is necessary to include the dorsal region in order to obtain a complete embryo, This is suggested by the fact that separation of the first two blastomeres can produce two complete half-size embryos if the first cleavage is medial, but often produces one complete half-size embryo plus one 'belly piece' if the first cleavage is frontal (Spemann, 1938).

In the blastula, the work of Nieuwkoop suggests that following division of the embryo along the egg axis only the prospective marginal zone (zone III) can form a more or less complete embryo in isolation. However, the combinations I + II + III and I + IV also form reasonably complete embryos, indicating that zone III can be re-formed from the extremes of the egg axis. These zones each comprise about a quarter of the volume of the egg, so presumably the size reduction is not in itself a sufficient reason for a failure of regulation, since we have previously seen that normally proportioned embryos can develop from quarter-sized fragments of fertilised eggs.

Several 'two-gradient' theories have been advanced in the past to explain various aspects of early amphibian development (Dalcq & Pasteels, 1937; Saxén & Toivonen, 1962). The following description is an extension of these and an attempt to summarise the results in a form which suggests something about the underlying mechanism of the interactions.

1. The blastula is partitioned into territories arbitrarily labelled 1–4 in the egg axis and 1′–4′ in the ventral–dorsal axis. Each territory is identified by two codings: e.g. (3,3′) and left and right sides are mirror images with identical codings.
2. If the embryo is fragmented or rearranged the codings become re-established across the space available. Territories can be halved in size in each linear dimension but cannot be further reduced.
3. The extreme codings are the highest present in the fragment. High values can generate lower ones but not vice versa.

4. By the commencement of gastrulation the codings begin to control the behaviour of the cells in each region.

This description is illustrated in Fig. 3.13. It will be noticed that there is one region of the blastula which has the highest codings in both axes and can, therefore, in principle, generate all the other values in the embryo. This region is called the organiser.

The organiser

No subject in embryology has been so misrepresented and misunderstood as the properties of the organiser. This is because there are several interactions involved in early amphibian development and a number of different assay procedures which can detect one or more of them at a time. In particular, the organiser should not be confused with 'mesodermal induction' which is the regionalisation along the egg axis discussed above, nor with 'neural induction' which is discussed below.

The properties of the organiser must be understood in the context of grafting experiments on the early gastrula, of which some of the most definitive are interspecies grafts carried out between *Triturus vulgaris* (=*taeniatus*) and *T. cristatus* (reviewed by Spemann, 1938). The former has more deeply pigmented eggs and so it is possible to identify at later stages which parts of the embryo are derived from the graft and which from the host. Three types of graft develop according to their new positions. These are dorsal ectoderm to ventral ectoderm, ventral ectoderm to dorsal ectoderm, and ectoderm to mesoderm. These results suggest that there is no irreversible determination of dorsal *versus* ventral ectoderm, and that, as in the blastula, it remains possible to vegetalise the ectoderm. They do not, however, show that these parts of the early gastrula are completely unspecified, or to use Spemann's term 'indifferent', indeed we have seen above that the entire ectodermal region is specified to become epidermis at this stage.

The original organiser graft is a graft of dorsal mesoderm from the early blastopore into the ventral marginal zone (Spemann & Mangold, 1924). In favourable cases this graft continues its invagination to form a second set of dorsal mesodermal structures, mainly notochord and prechordal plate, on the ventral side of the host. It also 'captures' a certain region of host tissue and 'dorsalises' this so that it becomes the outer part of the secondary mesodermal axis (somites, kidney, lateral plate) (Figs. 3.14 and 3.15). The subsequent events of neural and endodermal regionalisation follow from this, so that the end result can be the formation of a complete 'secondary' embryo on the ventral side, whose parts are arranged in a relation of mirror symmetry to those of the original embryo. The two salient characteristics of the organiser are first that its own state of determination is not affected by its new position, and secondly that it

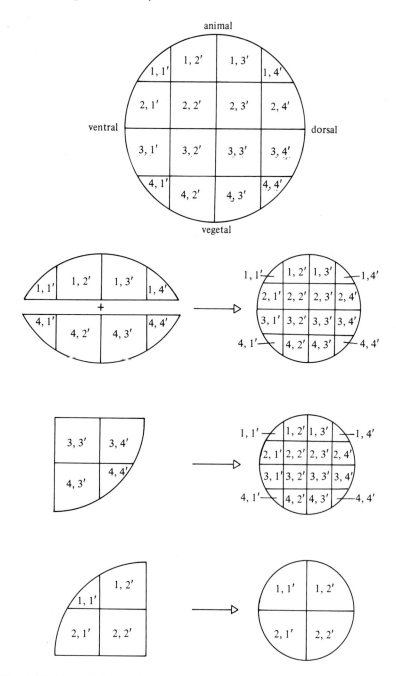

Fig. 3.13. Formal description of the regulative properties of the amphibian blastula. High codings can generate lower ones but not vice versa. A twofold reduction in linear dimensions is allowed for the size of each territory.

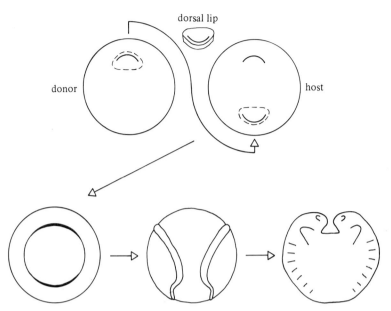

Fig. 3.14. The organiser graft. A piece of tissue from the dorsal marginal zone is grafted into the ventral marginal zone and induces the surrounding tissue to participate in the formation of a double-dorsal embryo.

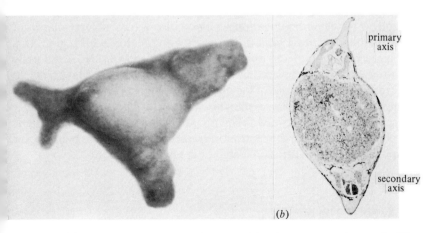

Fig. 3.15. (*a*) A double embryo of an axolotl arising from an organiser graft. (*b*) Transverse section through the trunk of another specimen (*Pleurodeles*). The mirror symmetry is quite apparent. (Author's photographs.)

can dorsalise the surrounding ventral mesoderm of the host to form a second embryonic axis.

An alternative method of performing the organiser graft, introduced by Mangold, involves the insertion of the graft into the blastocoel through a hole made in the ectoderm. This method is easier than implantation into the ventral marginal zone and much of the classical work on the organiser made use of it. Unfortunately it is difficult to control exactly where the graft ends up and so the result is a product of several processes which operate to different extents depending on the details of the experiment; namely vegetalisation of ectoderm in contact with the graft, dorsalisation of this induced mesoderm or of host mesoderm, self-differentiation of the graft, and, last but not least, neural inductions. More than any other single factor it is the complexity of this experiment which led to the idea that the organiser is an agent of neural induction rather than an agent of mesodermal dorsalisation.

Recent studies of the organiser have been made by Cooke (1972, 1973, 1979, 1981) using *Xenopus laevis*. The capture by the graft of a certain invagination territory is rapid, within 1–2 hours. It is always smaller than the invagination territory of the host organiser unless the host dorsal lip region is simultaneously removed. The number of cells in the mesoderm of double embryos is the same as that in control embryos, which implies that the ventral mesoderm is dorsalised without any extra cell division taking place. Finally, the organiser not only captures the ventral part of the host but also compresses the pattern of the host mesodermal structures in the dorsoventral axis, so the total number of distinctly determined zones in the double embryo is increased. Using explants cultured *in vitro*, Slack & Forman (1980) showed directly that the ventral marginal zone could be dorsalised by co-culture with organiser tissue. Explants which normally produced blood islands formed large masses of muscle after a certain period in contact with the dorsal lip region. It is thought that tissue from about 60 degrees on either side of the early blastopore has organiser activity (Bautzmann, 1926; Mayer, 1935; Minganti, 1949). In terms of the models discussed in Part III this evidence suggests that the organiser is the source region for a dorsoventral gradient responsible for the regional subdivision of the mesoderm. However, it is probably not a simple concentration gradient, the properties being better explained by one of the more complex gradient models.

Regulation in the gastrula

The regulative abilities of the early gastrula are less than those of the blastula, perhaps because there is now not much time left before large parts of the embryo become irreversibly determined. It is still possible to remove the ventral half of the early *Triturus* gastrula and obtain a reasonably well proportioned embryo from the dorsal half (Ruud & Spemann, 1922). However, defects in the dorsal region often lead to defects in the later embryo unless they are very small.

The nuclei of gastrulae are not as effective at supporting development of enucleated eggs as those from blastulae, but some will still do so. The consensus among nuclear transplantationists is that the ability to support egg development declines gradually but that it is possible to obtain a few positive results even from nuclei of adult differentiated cells. The decline is thought to be due to factors other than an irreversible reduction in potency.

Formation of the craniocaudal pattern in the course of invagination

On the dorsal side of the embryo the region destined to invaginate may extend 30 to 90 degrees (depending on species) from the blastopore. The fates of regions of this dorsal marginal zone are, from the blastopore towards the animal pole, first pharyngeal endoderm, then prechordal plate, then notochord, and finally the dorsal part of the tailbud (Okada & Hama, 1945). The effects of removing the dorsal lip region at different stages of gastrulation are consistent with this fate map. Removal of the dorsal lip from early gastrulae leads to deficiencies in the head region, although such defects often regulate; removal of dorsal lips from mid-gastrulae leads to deficiencies in the trunk region, and removal of dorsal lips from late gastrulae leads to deficiencies in the tail (Lehmann, 1926; Shen, 1937).

But the self-differentiation behaviour of small explants from this region at early gastrula stages is rather different from the fate map. The region closest to the blastopore tends to form notochord and neural tissue and the region further away tends to form epidermis and pigment cells (Holtfreter-Ban, 1965). The prospective trunk mesoderm only acquires a self-differentiation behaviour of notochord/neural type by the mid-gastrula (Kaneda & Hama, 1979). Specification of the dorsal mesoderm does not, therefore, correspond to its fate until after invagination.

When grafted to the ventral marginal zone, an early dorsal lip can, as we have seen, form a complete secondary embryo with head, trunk and tail. But a dorsal lip from a mid-gastrula grafted to the ventral marginal zone of an early gastrula will form only trunk and tail, and a dorsal lip from a late gastrula only tail (Spemann, 1931; Mayer, 1935). The dorsal lip itself contains both invaginated and uninvaginated tissue and it seems probable that the craniocaudal level of determination of each cohort of cells is achieved during invagination. This can be assayed by testing for neural induction activity by implantation into the blastocoel. In such experiments the dorsal marginal zone of the uninvaginated early gastrula induces tail structures while the dorsal mesoderm of the invaginated early gastrula induces head structures (Takaya, 1978).

Gastrulae treated with lithium chloride tend to develop into embryos that have axial defects (Lehmann, 1937; Bäckström, 1954). If early gastrulae are treated the defects are concentrated in the head while if later gastrulae are treated the defects occur in the head and also more posteriorly. The primary effects are certainly on the mesoderm but gaps arise in the central nervous system as a consequence of the failure of neural induction. Hence cyclopia and anophthalmia are common head defects.

All this evidence seems to suggest that the dorsal mesoderm acquires its craniocaudal state of determination on invagination.* Before invagination the self-differentiation behaviour is still governed by the signals in the blastula (Fig. 3.13). After invagination, the archenteron roof (as it has now become) is determined to become head, trunk or tail, and, as we shall see below, to induce corresponding neural structures from the overlying ectoderm. Future research on the mechanism of craniocaudal determination will probably focus on the changing sequence of microenvironments to which the invaginating cells are exposed.

Neural induction

In the original organiser graft of Spemann & Mangold (1924) one of the results, and the one on which most subsequent attention has concentrated, was that the graft induced a second nervous system from that part of the host ectoderm which came to lie above it at the end of gastrulation. In the absence of the graft, this tissue would have become ventral epidermis. Since it was also known that prospective epidermis and neural plate (dorsal and ventral ectoderm) could be exchanged in the gastrula and would each develop according to their new position, Spemann (1938) argued that the nervous system was formed in *normal* development as a result of an inductive signal from the archenteron roof. This interpretation was strengthened by the discovery of Holtfreter (1933) that the entire ectoderm could develop into epidermis if isolated from the mesoderm. This occurs in 'exogastrulae' which arise if embryos are cultured in stronger salt solution than usual; the morphogenetic movements work to exclude the mesoderm and endoderm from the ectoderm so that the ectoderm becomes isolated as an empty sac.

In 1932 it was shown that killed dorsal lips would induce some neural

---

* It should perhaps be mentioned that Spemann (1931) reported induction of head structures by late dorsal lips which is not compatible with this interpretation. However I have examined some of his slides held at the Hubrecht Laboratory and I am satisfied that this result can be explained by secondary interactions between the induced neural plate and that of the host.

tissue following implantation into the blastocoel (Bautzmann *et al.*, 1932). Because of the confusion between the organiser and the neural inductor properties of the dorsal lip this led to the famous 'gold rush' for the biochemical basis of the organiser. It was found that many tissues from adult animals or indeed many purified chemical substances would induce neural tubes or patches of neuroepithelium from gastrula ectoderm. After a period of confusion, it became clear that the ectoderm was fairly delicately balanced between epidermal and neural pathways and that a variety of stimuli could tip the balance one way or the other.

Regional specificity

One problem about much of the early work on neural induction was that the response consisted in the formation from gastrula ectoderm simply of nondescript patches of neuroepithelium sometimes rolled up into vesicles of neural tube. This made it unclear whether these reactions really resembled neural induction by the archenteron roof since the salient characteristic of the natural signal is regional specificity, different parts of the archenteron roof inducing different parts of the central nervous system. This specificity was shown by Mangold (1933) who implanted explants from different craniocaudal regions of archenteron roof into the blastocoel of host *Triturus* embryos, and also by Holtfreter (1936) who placed ectodermal explants on different regions of the surface of exogastrulae. Similar results were obtained by Horst (1948) in combinations of these explants with gastrula ectoderm in isolation.

This implies that several distinct signals are involved in neural induction and this is usually recognised in diagrams which attempt to account for the course of development in terms of chains of induction. However there is also evidence that there can be interactions within the neural plate itself. It has been known for a long time that explants of neural plate are themselves neural inductors. This is called 'homeogenetic induction' (induction of like by like), and was shown by Mangold (1933) to have some degree of regional specificity, in that prospective brain induced brain and prospective tail induced tail. A possible mechanism for homeogenetic induction is provided by the type of biochemical switch discussed in Chapter 10. Nieuwkoop (1952*a*, *b*, c) showed that if small folds of gastrula ectoderm are implanted into a neural plate, the basal parts of the folds develop into similar structures as are formed by their surroundings while the apical parts of the folds form more anterior structures that tend to be arranged in the normal sequence. So anterior neural plate could induce only forebrain, but posterior neural plate could induce a complete sequence of structures.

These indications that interactions can occur within the neural plate, although perhaps only in abnormal situations, suggest that the signals

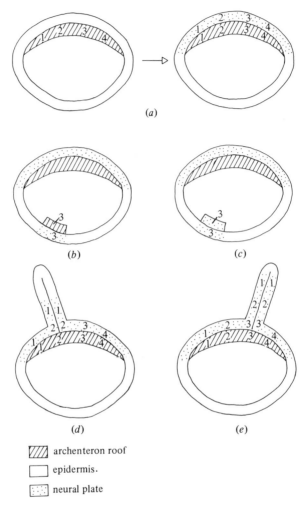

☐ archenteron roof

☐ epidermis.

☐ neural plate

Fig. 3.16. Neural induction. (*a*) In normal development the neural plate is induced by the archenteron roof. (*b*) The archenteron roof can induce neural tissue from prospective epidermis. (*c*) Neural plate can induce further neural tissue of the same regional specificity from prospective epidermis (homeogenetic induction). (*d*) and(*e*) Serially ordered structures are induced in ectodermal folds which were implanted into the neural plate.

from the archenteron roof may be representable as some sort of serial hierarchy as implied in Fig. 3.16, where they are shown as a series of numbers. This idea is implicit in the gradient models for neural induction put forward by Yamada and by Toivonen (Saxén & Toivonen, 1962), and possible mechanisms will be discussed further in Chapters 9 and 10.

Biochemical mechanism of neural induction

Biochemical studies on neural induction have been bedevilled by the complexity of the situation *in vivo*. As has been noted above this makes the relevance of some of the assays which have been used for active substances rather questionable. It seems to be agreed that a 'neuralising factor' (N factor) can be extracted from a number of adult tissues, such as liver, and that it is heat stable. On its own it induces forebrain structures ('archencephalic induction'). The M factor, which has already been discussed in connection with the regionalisation of the blastula, will induce notochord, somite and neural tube ('spinocaudal induction'). When N and M are applied together, structures appropriate to all craniocaudal levels may be induced (Saxén & Toivonen, 1962; Tiedemann, 1976; Toivonen, 1978).

Experiments have been carried out in which the inductor and the reacting tissue are separated by various types of filter. Saxén (1961) showed that neuralisation could take place across a Millipore filter. Tarin, Toivonen & Saxén (1973) showed that it could take place across Nucleopore filters of small pore size (down to 0.1 μm) but not across Cellophane membranes. In these experiments dorsal blastopore lip was used as the inducer so presumably the signal involved is the natural one. It is believed that 0.1 μm Nucleopore membranes cannot be penetrated by cytoplasmic processes and so the implication is that the natural signal, as well as various unnatural ones, can be transmitted in the form of extracellular diffusible molecule(s).

Studies on the electrophysiology of the neural plate by Warner (1973) suggest that neural differentiation in the axolotl is under way as early as the neural plate stage. The resting membrane potentials rise from about -15 mV characteristic of ectoderm to about -40 mV at this time. The region in the centre of the 'keyhole' of the neural plate develops the highest potential and since the whole neural plate is electrically coupled this implies that there is a current flow towards this region. Whether these early events are causes or consequences of the regionalisation of the neural plate is not known, although there is evidence that a functional sodium pump is required for differentiation of neurons *in vitro* (Messenger & Warner, 1979).

Somitogenesis

During neurulation, the mesoderm on either side of the neural tube becomes segmented to form a row of somites. These later form the myotomes, the vertebrae, and part of the dermis of the skin. The cell movements which lead to segmentation differ somewhat from species to species but in all cases the segmentation occurs in a craniocaudal sequence and is temporally correlated with events occurring elsewhere in the body. It is unlikely that the size of somites depends on a cell counting process since haploid embryos have been shown to contain normal-sized somites each containing twice the normal number of cells (Hamilton, 1969). When embryos are transected at a level posterior to the last-formed somite both halves continue segmentation at the normal tempo and end up with the number of somites which they would have developed in the intact animal (Deuchar & Burgess, 1967). To this extent the segmentation pattern is a mosaic from the end of gastrulation. However, although it is not possible to alter the fate map it is possible to disturb the arrangement of cells by exposing the embryos to brief temperature shocks at 37°C. During the neurula and tailbud stages the somites which become deranged are those which are formed a few hours after the temperature shock, the exact time depending on the species (Fig. 3.17; and Pearson & Elsdale, 1979). Therefore there seems to be a craniocaudal progression of temperature sensitivity which runs in advance of visible segmentation. This process is also unaffected by transection of the embryo and thus presumably does not depend on any long-range cellular interactions. The authors call the temperature-sensitive event 'somite determination' and argue that the cells must be able to recover from the temperature shock after a few hours since only a short length of the file is deranged and the somites posterior to the lesion are normal.

Temperature shocks given to late blastulae or ealy gastrulae produce more serious defects in segmentation which may involve the whole somite file or any part of it (Elsdale & Pearson, 1979). This is interpreted as being the stage at which the general craniocaudal pattern is established in the mesodermal mantle. Shocks given to the late gastrula are without any effect at all, and this refractory period between two periods of sensitivity indicates that somite determination involves two distinct processes: one early and the other just preceding visible segmentation. These results are compatible with the 'clock and wavefront' model for the determination of repeating structures which will be discussed in Chapter 10.

The endoderm

The regionalisation of the endoderm has not been the most popular of research topics but the results we have suggest that the territories become determined by the middle neurula under the influence of the adjacent mesoderm.

A fate map of the early neurula endoderm was produced by Balinsky (1947) and used to examine the results of interchanging prospective regions for stomach and liver (Balinsky, 1948). Normal embryos arose from these grafts, indicating that determination had not yet occurred. The question was later examined by Okada using explants (Okada, 1953,

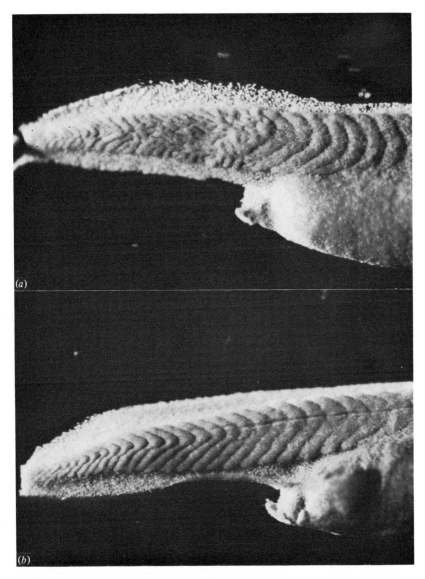

Fig. 3.17. Localised disruption of somite segmentation in *Rana* embryos arising from a brief temperature shock: (*a*) experimental, (*b*) control. (Photographs kindly provided by Dr T. Elsdale.)

1957, 1960). He found that gastrula endoderm alone would not differentiate *in vitro*. Prospective pharyngeal endoderm would form pharynx in combination with archenteron roof tissue, and intestine in combination with lateral plate mesoderm. Prospective gastric and intestinal endoderm

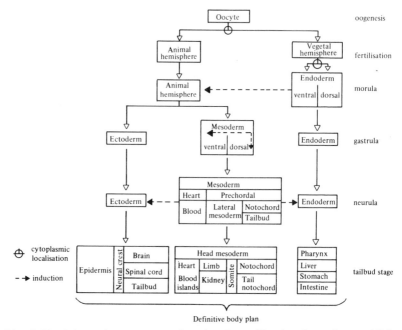

Fig. 3.18. Schematic summary of regional specification in early amphibian development.

would form branchial or oesophageal structures in the presence of head mesoderm, and intestinal structures in the presence of lateral plate mesoderm.

Summary of determinative events in the early amphibian embryo

The results which have been described here may perhaps be brought together and summarised as follows (Fig. 3.18). Animal–vegetal polarity is established during oogenesis and dorsoventral polarity is established after fertilisation. Both of these apparently involved cytoplasmic localisation but we know nothing about the molecules which we presume to be segregated to the different regions.

The first inductive interaction occurs along the egg axis and results in the formation of the mesoderm. The process is mimicked *in vitro* by the action of the M factor on isolated ectoderm. As soon as the mesoderm is formed the dorsal and ventral regions seem to be differently specified. The dorsal region is the organiser and emits an inductive signal which regionalises the mesoderm in the dorsoventral and perhaps also in the craniocaudal axes before and during gastrulation.

In the gastrula and neurula stages the neural plate is induced by the action of the mesoderm on the overlying ectoderm. The neural plate is formed as an array of territories committed to become different parts of the nervous system. The endoderm is similarly regionalised under the influence of the mesoderm to form different parts of the viscera.

Up to the time at which the basic body plan has become determined we therefore have at least two decisions involving cytoplasmic localisation and at least five inductive signals, some of which have multiple outcomes and may be complex signals involving many biochemical substances.

Although the signals and responses involved in regional specification are much better characterised in early amphibian embryos than for other types of embryo we still have very little idea about the underlying mechanisms. We do not know either the identities of the chemical substances involved or the dynamical properties of the systems which must underlie such phenomena as proportion regulation or homeogenetic induction. The explanations of classical workers in this field must now be regarded as problems for solution.

General references

Spemann, H. (1938). *Embryonic Development and Induction.* (Reprinted in 1967 by Hafner, New York.)

Nakamura, O. & Toivonen, S. (eds.) (1978). *Organizer: A Milestone of a Half Century from Spemann.* Elsevier/North-Holland, Amsterdam.

Gerhart, J. C. (1981). Mechanisms regulating pattern formation in the amphibian egg and early embryo. In *Biological Regulation and Development*, vol. 2, ed. R. Goldberger, pp. 133–316. Plenum Press, New York.

4

Insect embryos and the genetic program
of early development

At first sight insect eggs do not seem very suitable for experimental manipulation. They are smaller than amphibian eggs and they are surrounded by a tough impermeable chorion without which they cannot survive. However, insects now have a special position in embryology because of the possibilities for the genetic analysis of early developmental decisions. The fruit fly, *Drosophila melanogaster*, has been intensively used in genetic research for 70 years and over this time a large number of mutants have been collected and accurate genetic and cytological maps of the chromosomes have been compiled. Because of the small size of *Drosophila* and because of its short life cycle of 2 weeks, it is possible to screen large numbers of eggs for lethal mutations affecting the basic body plan, and this has led to a degree of detail in understanding the genetic program of development which does not exist for any other kind of animal. In this chapter these results will be set in the context of embryological results obtained by micromanipulative experiments on other species of insect which are more suitable for such work on account of their larger size and slower development.

All adult and larval insects are built up of an anteroposterior sequence of segments which fall into the three principal body regions of head, thorax and abdomen; the basic segmental arrangement is shown in Fig. 4.1, which depicts a late germ band stage of the beetle *Tenebrio*. The head is not visibly segmented in adult insects but in the embryo it may consist of as many as six segments: three procephalic and three gnathal, the gnathal segments bearing leg-bud-like appendages which later become the mouthparts. The middle part of the body is the thorax, which always consists of three segments: the prothorax, mesothorax and metathorax. All of these bear a pair of legs on the ventral side and the meso- and metathorax also bear a pair of wings on the dorsal side. The number of abdominal segments varies with species but is usually in the range eight to eleven.

The species discussed in this chapter are listed in Table 4.1. In the last column the length of the germ anlage is described as long, intermediate or short (Krause, 1939). The germ anlage is the condensation of cells which

Table 4.1. *Insect species mentioned in this chapter*

Name	Common name	Taxon	Time to hatch (days)	Length of germ anlage
Tachycines	Camel cricket	Hemimetabola: Orthoptera	40	Short
Platycnemis	Dragonfly	Hemimetabola: Odonata	18	Intermediate
Oncopeltus	Milkweed bug	Hemimetabola: Hemiptera	7	Intermediate
Euscelis	Leafhopper	Hemimetabola: Homoptera	17	Intermediate
Bruchidius	Bean weevil	Holometabola: Coleoptera	9	Long
Bombyx	Silk moth	Holometabola: Lepidoptera	14	Long
Smittia	Midge	Holometabola: Diptera	3½	Long
Drosophila	Fruit fly	Holometabola: Diptera	1	Long

The times to hatch are approximate and refer to ambient temperatures.

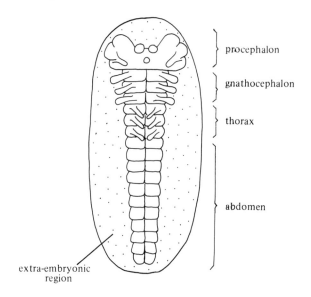

procephalon

gnathocephalon

thorax

abdomen

extra-embryonic
region

Fig. 4.1. The general body plan of an insect is exemplified by this diagram of a late germ band stage embryo of the beetle *Tenebrio*. (After Ullman, 1964.)

gives rise to the embryo, as opposed to the extra-embryonic parts, and the terms refer to the proportion of the larval body plan represented on the fate map of the germ anlage. As we shall see this is associated with differences in descriptive embryology between the types. As indicated in the table, insects are divided into two main taxa: the Hemimetabola and Holometabola (now often called Exopterygota and Endopterygota by zoologists). There is also a group of primitive wingless insects which does not concern us here. The Hemimetabola hatch from their eggs as larvae which are fairly similar in morphology to the adults and are called nymphs. As they grow the nymphs pass through a number of moults in each of which the old cuticle is shed and a slightly more adult-like nymph emerges. The Holometabola are so called to indicate by contrast their abrupt and complete metamorphosis. Usually the larva which hatches from the egg is quite different in structure from the adult. It grows and passes through a definite number of moults before becoming a resting stage called a pupa in which the body is remoulded to form the adult. During this metamorphosis some larval structures are retained, some are reorganised and some are replaced by new structures which develop from imaginal buds. *Drosophila* undergoes an especially drastic metamorphosis in which the entire cuticle of the adult is formed from such buds (Table 4.2). Of these buds the imaginal discs grow extensively during larval life while the abdominal histoblasts are mitotically quiescent during larval life but grow in the pupa. The imaginal discs have been subjected to

Table 4.2. *Imaginal reserve buds in the* Drosophila *larva*

Bud	Becomes
Clypeolabral disc	Clypeolabrum
Eye–antennal disc	Eyes, antennae, rest of head
Labial disc	Proboscis
First leg disc	First leg + ventral prothorax
Wing disc	Wing + dorsal mesothorax
Second leg disc	Second leg + ventral mesothorax
Haltere disc	Haltere + dorsal metathorax
Third leg disc	Third leg + ventral metathorax
Dorsal abdominal histoblasts	Tergites
Ventral abdominal histoblasts	Sternites
Genital disc	Genital apparatus

The genital disc is bilaterally symmetrical and lies in the midline. All the others are paired with one member on the right and the other on the left side. Some authors would also include a dorsal prothoracic disc forming the humerus.

a good deal of experimental work in their own right in connection with clonal analysis, regeneration and transdetermination (reviewed by Ursprung & Nöthiger, 1972), but these topics will not be discussed here since they lie outside the scope of the present book. For the moment the important thing to notice is that the original rudiments for the imaginal buds are laid down at a very early stage of embryonic development and therefore they make up an important part of the *larval* body plan even though they remain undifferentiated until pupation.

Normal development

The normal development of a generalised insect is depicted in Fig. 4.2. Fertilisation occurs through a pre-existing hole in the chorion called the micropyle. After fusion of the male and female pronuclei the zygote nucleus undergoes a number of synchronous divisions. Although this phase is known as cleavage, no cell membranes are formed and the embryo consists of a syncytium in which nuclei, each accompanied by a little cytoplasm as so-called cleavage energids, lie free in the yolk. As division proceeds the nuclei spread throughout the egg, and at a certain stage most of them migrate to the periphery to form the syncytial blastoderm, while a few remain in the yolky interior and later become vitellophages. After a few more divisions, cell membranes develop as ingrowths from the plasma membrane around each nucleus. This is the

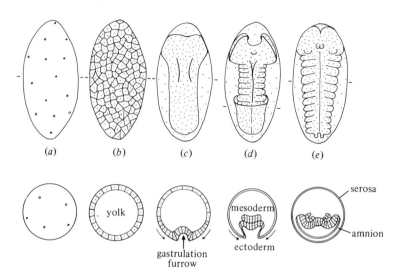

Fig. 4.2. Embryonic development of a generalised insect. The lower row of diagrams shows transverse sections through the levels indicated on the upper row. (*a*) Cleavage, (*b*) blastoderm, (*c*) germ anlage, beginning of gastrulation, (*d*) germ band, formation of embryonic membranes, (*e*) late germ band stage. (Adapted from Sander, 1976.)

cellular blastoderm stage at which the embryo consists of a uniform epithelium around the outer surface of the egg surrounding an acellular yolky interior.

Following this there is a concentration of cells on the ventral side, often occupying only part of the egg length. This is the germ anlage which is the rudiment for the embryo proper, the thin layer of epithelium over the remaining surface later becoming the extra-embryonic serosa.

At some early stage, depending on species, the germ cells are formed. Many insect eggs contain a special region of granular cytoplasm near the posterior pole, and it is the cells which inherit this cytoplasm (pole cells) which become, or at least include, the primordial germ cells. They later migrate anteriorly and enter the gonads.

In the next phase of development the germ anlage develops into a multilayered germ band and a number of different things happen which may occur in different sequence depending on species. Note that the terms 'germ anlage' and 'germ band' refer to the whole embryo and should not be confused with the germ cells which are the prospective reproductive cells.

In long-germ insects the sequence of visible segmentation usually proceeds from anterior to posterior and rudiments for the whole length of the body are established quite quickly. The early germ band stretches itself so that the posterior end bends over to the dorsal side. In short-germ insects only the rudiment for the head is present and the segments of thorax and abdomen are produced over a long period of time from a posterior growth zone. In insects with germ anlage of intermediate length, segmentation often commences in the thoracic region and proceeds from there both anteriorly and posteriorly, the abdomen being formed by a growth zone.

The morphogenetic movements of this phase involve the formation of the mesoderm by a mid-ventral invagination and of the endoderm by terminal invaginations. Gastrulation proper consists of the ingression, or invagination, of the mid-ventral strip of prospective mesoderm destined to become body musculature, gut wall, gonads and blood cells. It becomes segmented into 'somites' in the same sequence as the ectoderm. The gut arises from two regions. The stomodaeum and anterior mid-gut invaginate from the anterior end and the proctodaeum and posterior mid-gut invaginate from the posterior end. During gastrulation a number of neuroblasts differentiate in the ectoderm adjoining the ventral furrow. They sink in and later become the segmental ganglia of the ventral nerve cord. At about the same time the extra-embryonic ectoderm at either end of the embryo folds over it and the two folds fuse leaving the embryo covered by two membranes: the outer serosa and the inner amnion.

By the time that all of these events have occurred we have reached the stage of late germ band and all of the principal parts of the larval body are in their definitive positions. At this stage all insect species reveal their close affinity in basic structure, although they may diverge again on their way to the larval stage. In later development the circumference of the body is completed by a dorsal closure process in which the lateral edges of the ectoderm and mesoderm move up and join in the dorsal mid-line to enclose the gut.

The Hemimetabola may undergo some additional morphogenetic movements called anatrepsis and katatrepsis. Anatrepsis is an effective head to tail inversion of the germ band in the egg case, while katatrepsis is the return movement. Such movements can confuse the results of embryological experiments for the uninitiated because for a time the parts lie in the opposite order to that in which they were laid down.

Because the development of *Drosophila* is rather different from the insect norm it is perhaps worth noting some of its special features (Sonnenblick, 1950; Poulson, 1950). Development is very fast and the larvae hatch after 22 hours at 23°C compared with the days or weeks taken by other species (Table 4.1). The larva has no legs, its head is tucked away in the interior and it has three thoracic and eight abdominal segments. The early embryonic stages are depicted in Fig. 4.3.

Nine synchronous nuclear divisions occur in the cleavage phase after which most of the nuclei migrate to the periphery to form the syncytial blastoderm. The pole cells are formed very early, at about 1.5 hours.

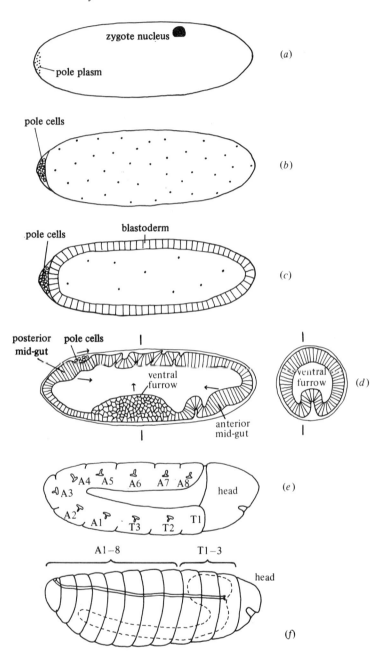

Fig. 4.3. Embryonic development of *Drosophila melanogaster*. T1–3, thoracic segments; A1–8, abdominal segments. (*a*) Zygote, (*b*) cleavage, (*c*) cellular blastoderm, (*d*) gastrulation, (*e*) maximum germ band extension, (*f*) after dorsal closure.

After four more nuclear divisions, at about 3 hours, the cellular blastoderm is formed. At this stage there are about 6000 surface cells (Turner
& Mahowald, 1976), 1000 yolk nuclei and 30–50 pole cells. There is then
a pause in mitotic activity for over an hour.

The columnar epithelium of the blastoderm is quite thick and most of it
is destined to become the embryo itself, only a thin dorsal strip
becoming the amnion and serosa. Gastrulation commences at 3.5 hours
with a deep ventral furrow. Neuroblasts become visible at 4–4.5 hours and
their arrangement constitutes the first visible indication of segmentation.
At this time there is one more round of mitosis in the ectoderm,
subsequent mitoses being largely confined to the internal tissues.

Segmentation in the mesoderm is apparent by 5 hours, and in the
ectoderm at 6–7 hours by the appearance of paired tracheal pits. During
gastrulation the elongation of the germ band drives the posterior end
along with the pole cells round the dorsal side of the egg. It returns before
dorsal closure at 10–11 hours.

The formation of the gut is fairly normal but that of the head is unusual.
It is delineated at 3 hours 45 minutes by a groove called the cephalic
furrow. The head mesoderm does not become visibly segmented while
three transient segments appear in the ectoderm. At the time of dorsal
closure the head 'involutes' into the interior and is therefore scarcely
represented on the exterior of the larva.

The imaginal discs and abdominal histoblasts are present as small nests
of cells in the first instar larva. It is not really known at what time the
rudiments become established in the embryo but evidence from clonal
analysis which will be discussed below indicates that it may be before the
time of the first post-blastoderm mitosis at 4–4.5 hours.

Genetics

Genetic arguments are notoriously difficult to follow because of the
apparent obscurity of some of the terminology. In this chapter a number
of terms will be used which are defined below; some of them are not the
normal forms used in *Drosophila* genetics but they should at least be
generally comprehensible.

The normal allele at each locus is called the wild type and is designated
by the symbol '+'. Usually each locus is present in two copies: one on the
maternally derived chromosome and one on the paternally derived
chromosome. If both loci are occupied by the same allele then the fly is
homozygous. If one is occupied by the wild type and the other by the
mutant allele it is heterozygous. If only one copy of the locus is present
because it is on the X chromosome in the male or because it has been
deleted from the other chromosome then the fly is hemizygous. A mutant
allele is recessive if it does not influence the phenotype in the presence of

the wild type allele, and dominant if it does express a phenotype in the presence of the wild type allele. Recessive mutations are usually thought to arise from inactivation of structural genes which should normally produce a protein product, and only when both copies of the gene are inactive (homozygote or hemizygote) will the product be absent and the effect of its absence be apparent. Not all mutant alleles will be completely inactive but the genetic analysis is facilitated by using those alleles which abolish the function of the gene completely, and it is now known that these are often deletions or insertions of stretches of DNA.

Dominant mutations may also arise from inactivation of structural genes in cases where a 50% reduction in the amount produced is sufficient to cause an effect. In such cases one would expect a stronger version of its phenotype in a homozygote than in a heterozygote, and this is in fact quite common among pattern mutants affecting the body plan. Often a dominant mutation will be lethal in the homozygous form because the phenotype is more extreme. On the other hand, dominant mutations which show the same phenotype in homozygotes as heterozygotes are usually assumed to be regulatory defects which cause an inappropriate activation of some different, structural, gene. Dominant mutations are usually symbolised by an initial capital letter and recessive mutations by an initial lower-case letter. However, pattern mutants are often named for some phenotypic feature which is not the one under consideration and so this convention is not always followed here.

The effects of mutations affecting pattern may be zygotic or maternal. If the effect is zygotic then the embryo displays the phenotype appropriate to its own genotype; if it is maternal then the embryo displays the phenotype appropriate to the genotype of the mother. Maternal effects are assumed to be due to faults in the composition or cytoarchitecture of the egg arising during oogenesis. Perhaps surprisingly, most of the known mutations affecting the body plan of *Drosophila* have zygotic effects, which implies that the embryonic genome is active by the time the principal decisions are made.

Fate maps

Vital staining and other direct methods of marking have not been applied to insect eggs because of the difficulty of penetrating the chorion. The fate maps which do exist have been based on indirect evidence: attempts to retrace the gastrulation movements with the aid of histological sections, the study of defects caused by localised irradiation, and gynandromorph maps.

For reasons which will become clear later it is not thought that early cleavage nuclei have definable fates. In other words early nuclei in the same position in different individual eggs will probably contribute to

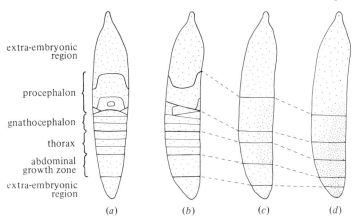

extra-embryonic region

procephalon

gnathocephalon

thorax

abdominal growth zone

extra-embryonic region

(a)   (b)   (c)   (d)

Fig. 4.4. Fate map of *Platycnemis* according to Seidel (1935). (*a*) and (*b*) blastoderm stage, (*c*) germ aggregation stage, (*d*) germ anlage stage.

different sets of structures. We are therefore concerned with the fate maps for stages between the syncytial blastoderm, after which random mixing occurs only on a small scale, and the late germ band, by which time the definitive rudiments can be visualised.

Seidel (1935) conducted an exhaustive study of defects induced by localised ultraviolet (UV) irradiation of embryos of the dragonfly *Platycnemis*, an insect of intermediate-length germ anlage. As Seidel himself pointed out, a defect map is not really the same as a fate map because the marking method may perturb normal development. In particular if a signalling centre is destroyed then the pattern may be disturbed outside the irradiated region, and conversely if the signalling centres are not damaged then the dead cells may be replaced by healthy ones which move in and assume their fates.

However, if we set these problems aside, Seidel's fate maps show three things quite clearly (Fig. 4.4). Firstly, the anteroposterior sequence of prospective regions in the ectoderm is maintained throughout development, which implies that there are no cryptic movements or mixing events. Secondly, the fate map starts off by covering most of the egg at the blastoderm stage and then contracts as the germ band is formed, presumably because of an aggregation of cells in the ventroposterior region. Thirdly, although the abdominal growth zone is present on the fate maps, the individual abdominal segments are not. So presumably there is some indeterminacy between individuals regarding which particular cells in the growth zone will contribute to which segment. A more recent but conceptually similar study was conducted on blastoderm-stage *Drosophila* embryos using a UV laser microbeam (Lohs-Schardin, Cremer & Nüsslein-Volhard, 1979). It was shown that prospective regions existed for all the larval segments and that they were arranged in a longitudinal sequence across about 50% of the egg length. So as in *Platycnemis*, the sequence of segments seems to be conserved throughout development, but since *Drosophila* is a long-germ species all of its segments are represented on the fate map as early as the blastoderm stage.

Like many things in developmental biology, the gynandromorph technique has caused great interest in the last 10 years but was actually discovered a long time ago. The basic method was applied to *Drosophila simulans* by Sturtevant (1929), revived by Garcia-Bellido & Merriam (1969) and then applied on a large scale to *Drosophila melanogaster* (review by Janning, 1978). Gynandromorphs are a type of genetic mosaic, or creature composed of cells with different genotypes. Genetic mosaics should not of course be confused with the 'mosaic behaviour' in grafting or isolation experiments which was discussed in Chapter 2. Gynandromorphs arise by the elimination of a chromosome from one nucleus after an early cleavage, preferably the first, which allows a recessive allele on the remaining chromosome of the pair to be expressed in all the cells descended from the monosomic nucleus. The mosaicism can be visualised if the mutation alters cuticular pigmentation or some enzyme which is normally expressed in most tissues but is not essential for survival (Fig. 4.5). It is believed that the first cleavage has a near-random orientation and therefore the boundary between the two genotypes is different in different individuals. Clearly, the further apart are two prospective regions on the fate map, the more likely are the resulting structures to be of different genotypes. So the fraction: [number of cases different]/[total mosaics] represents a distance on the fate map, and following a suggestion of Hotta & Benzer (1972) a value of 1% is called one 'sturt' after Sturtevant. After a number of such distances have been established, the points can be positioned on a two-dimensional surface by triangulation.

The chromosome used is usually the 'ring-X' (symbolised by R), an X chromosome which is frequently lost spontaneously at early cleavages. In female embryos which are originally XR (one normal X and one ring-X) loss of R from one of the products of the first nuclear division yields a fly which is half XR and half XO, the latter being a male genotype equivalent to XY except that the sperm are non-functional. The name gynandromorph in fact means a creature composed of both female and male parts. In Fig. 4.6 are shown two fate maps: one inferred from histological observation (Poulson, 1950), and the other a gynandromorph map for the larval structures and the imaginal discs (Janning, 1978). Poulson's map refers to the cellular blastoderm stage and since the gynandromorph map is clearly fairly similar it is usually assumed that it too refers to this stage. From a theoretical point of view we can say that it should be no earlier than the syncytial blastoderm since this is the time that the nuclei move out to label the peripheral volume elements of the egg.

As in the defect and the observational maps, the anteroposterior sequence of ectodermal prospective regions is maintained throughout

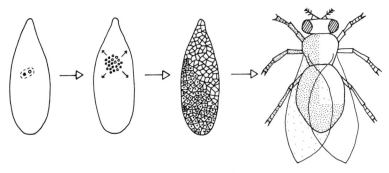

Fig. 4.5. The formation of a gynandromorph in *Drosophila*. A chromosome is lost from one of the first two cleavage nuclei. Each nucleus founds a clone and the boundary between the clones is randomly oriented with respect to the egg axis. The closer together are two points on the blastoderm, the more likely they are to be of the same genotype.

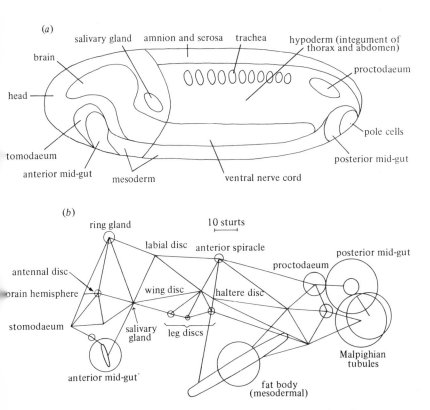

(a)

salivary gland   amnion and serosa   trachea   hypoderm (integument of thorax and abdomen)

brain

proctodaeum

head

tomodaeum

pole cells

anterior mid-gut   mesoderm   ventral nerve cord   posterior mid-gut

(b)

ring gland   10 sturts

labial disc   anterior spiracle   posterior mid-gut

antennal disc   proctodaeum

brain hemisphere   wing disc   haltere disc

stomodaeum   salivary gland   leg discs

Malpighian tubules

anterior mid-gut

fat body (mesodermal)

Fig. 4.6. Fate maps of the *Drosophila* blastoderm with respect to larval structures. (a) is constructed from morphological observations (adapted from Poulson, 1950), and (b) is constructed from the analysis of gynandromorphs (after Janning, 1978).

development. The pole cells map to the posterior, and the endodermal and mesodermal structures to the positions they would be expected to occupy before gastrulation.

The gynandromorph technique is very clever but is only as accurate as the assumptions on which it is based. One of these is that there must be no selection and no preferred developmental pathways between the two clones. In fact there is a tendency for structures to be mutant type less frequently than wild type, and this might be due to several causes including differences in division rate during cleavage and loss of the unstable chromosome at a division later than the first (Zalokar, Erk & Santamaria, 1980).

Primordial cell numbers

The question of the number of cells which contribute to a subsequently formed structure is one which has greatly interested developmental geneticists, mainly because the methods of clonal analysis provide the possibility of making measurements. As was mentioned in Chapter 2 an estimate has a very different significance when it is made before and after the stage of determination. After determination the number of cells in a rudiment must either stay the same or increase (assuming that there is no cell death). However, before determination the occurrence of local cell mixing can mean that the pool of cells which may, statistically, contribute to the structure in question is larger than the number which actually does so in each individual embryo. Some authors have been puzzled by the fact that the number of progenitor cells sometimes seems to shrink as determination takes place, but this should not be surprising to those who understand the true nature of a fate map.

The number of progenitor cells at the blastoderm stage can be estimated by finding the frequency with which the adult structure is internally mosaic in gynandromorphs. Clearly a single-cell structure such as a bristle must be of one genotype or the other, but an extended structure such as a wing will often be mosaic, and the frequency will depend on the size and the shape of the patch of prospective cells in the blastoderm. The *Drosophila* blastoderm consists of about 6000 cells, which implies a conversion factor of about 2 sturts = 1 cell diameter. A rod-shaped prospective region will have a frequency of mosaicism equal to its length in sturts, and a circular prospective region can be shown to have a frequency of mosaicism of the diameter $\times$ $\pi/2$ (Wieschaus & Gehring, 1976b). Estimates of cell numbers in the regions of the blastoderm which may contribute to certain imaginal structures are given in the first column of Table 4.3.

Gynandromorphs can also be used to estimate the cell number at determination time. This is done by finding the smallest possible patch of

Table 4.3. *Estimates of primordial cell numbers in prospective regions and rudiments of imaginal buds of Drosophila*

Structure	Estimate from % mosaicism in gynandromorphs	Estimate from minimum patch size in gynandromorphs	Estimate from size of clones induced by irradiation at blastoderm	Estimate by direct counts on first instar larvae
Wing disc	59	40	6	38
Leg disc	80	20	8	40
Eye–antennal disc	148	33	6	77
Tergite histoblast	10	9	4	20

Data simplified from Merriam (1978).

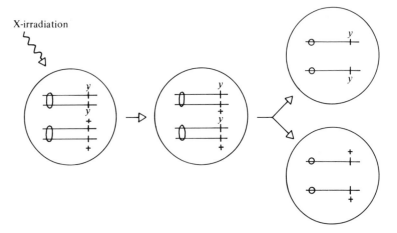

X-irradiation

Fig. 4.7. Creation of a marked clone by somatic crossing over. Every cell in the embryo is initially heterozygous for *yellow*. If crossing over occurs between strands of homologous chromosomes then two homozygous clones will be generated by the next mitosis, one +/+ and one *y/y*.

one genotype and dividing this into the total area of cuticle in the structure. The idea is that at least one cell of the genotype concerned must be incorporated into the rudiment to obtain mosaicism and if there are, say, 10 cells altogether and all of them grow equally in later development, then we would expect about 10% of the final structure to be marked. This method is inherently rather inaccurate since the significant observations come from a small minority of flies, but it does give an order-of-magnitude estimate and, as shown in the second column of Table 4.3, the numbers may be smaller than the cell numbers in the prospective region at the blastoderm stage.

A quite different method of making mosaic flies, and one which can also be used to estimate primordial cell numbers, is that of X-ray-induced mitotic recombination (Becker, 1957; Steiner, 1976; Wieschaus & Gehring, 1976a; Lawrence & Morata, 1977). Irradiation at a particular stage can cause mitotic recombination in a small proportion of the cells, and if the embryo is heterozygous for a recessive marker then such a recombination event will, after the next cell division, produce a marked double-recessive clone and an invisible + + clone (Fig. 4.7). If these clones grow at the same rate as the heterozygous background, and if there were *n* progenitor cells at the time of irradiation, then each should eventually occupy $1/n$ of the area of the structure.

In principle, the advantage of this technique is that clones can be induced at any time during development. So not only can primordial cell numbers be estimated at different stages but also it is possible to make estimates of times of determination on the basis that no clonal restriction means no determination (as discussed in Chapter 2). Lawrence (1973) carried out an analysis of X-ray-induced clones in the milkweed bug *Oncopeltus*. He found that some clones in the abdominal cuticle of the adult could span segment boundaries if they were induced before the germ anlage stage and that there was a rapid rise in the proportion of single-segment clones if irradiation was carried out after the syncytial blastoderm stage. In this species the germ anlage stage is reached about 24 hours after fertilisation and the first thoracic segments become visible by 36 hours, so presumably the determination of segment boundaries occurs during this period. Unfortunately the possibilities of clonal analysis cannot fully be realised in *Drosophila* because there is only a brief period in early development when the required radiation doses can be administered without killing the embryos. This is around the blastoderm stage when the nuclei have already undergone 13 of the 15 or so divisions which take place in the ectoderm in embryonic life. Histological observations made by Madhavan & Schneiderman (1977) suggest that the fourteenth mitosis occurs at 4–4.5 hours after fertilisation, which is the time of gastrulation. Since the marked clone is not generated until the first cell division after recombination, there would be very little chance of a clone spanning a segment boundary in the cuticle even if determination occurred after the blastoderm stage. For this reason there is still uncertainty about the timing of segment determination and of other events during early *Drosophila* development.

Estimates of primordial cell numbers in the prospective regions of imaginal buds during the radiation window of *Drosophila* are given in column 3 of Table 4.3. It can be seen that these are by no means the same as estimates from the gynandromorph percentage mosaicism method and these differences should remind us that despite the elegance of these techniques they are only as accurate as the assumptions upon which they are based. With respect to determination, no stage has been found at which X-ray-induced clones cross segment boundaries, or even the anteroposterior compartment boundary which exists in the thoracic imaginal discs, so it is often thought that these boundaries exist by gastrulation time. However, we cannot be sure of this. As emphasised in Chapter 2, clonal restriction does not prove determination. Non-restriction does prove non-determination but the conditions for obtaining non-restriction in the present case are very stringent. Firstly, it is necessary that there is enough random mixing of cells to give a marked clone a good chance of contributing to more than one segment. Secondly, it is necessary to wait for two cell divisions after irradiation: one to generate the clone and another to make it a two-cell clone which is the minimum size of clone which can possibly span two determined regions. Clones have been obtained which cross between wing and mesothoracic leg, although they are restricted to anterior or posterior compartments of both structures. It would seem from this that the distinction between dorsal and ventral disc rudiments is made later than that between segments or between anterior and posterior compartments of the thoracic discs, although exactly when it is made we do not know.

Determination of nuclei and properties of the pole plasm

Despite the small size of insect eggs, a number of transplantation experiments have been carried out on early stages of *Drosophila* and have yielded some of the most direct evidence on determinative decisions.

Nuclear transplantations between cleavage stages are of two types: transplantation into fertilised eggs and transplantation into unfertilised eggs. The former tests for the possibility of any type of reprogramming of the donor nucleus, while the latter tests for reprogramming specifically to the state of the normal zygote nucleus. The experiments are always conducted using donor nuclei of different genetic constitution from the host so that structures formed from the donor can be identified at a later stage. Since these have usually been adult cuticular markers (e.g. 'yellow', 'multiple wing hairs') it is necessary to rear the hosts not just to the larval stage but to the adult stage in order to obtain results.

Nuclei from cleavage, blastoderm and gastrula stages injected into the opposite end of a fertilised egg relative to their position of origin can become integrated into the host and produce patches of marked tissue in the adult corresponding to their new position. The experiment works less well with unfertilised eggs but a small number of fertile flies of donor genotype have been obtained after injection of blastoderm nuclei. This seems to show that the nuclei are not subject to any irreversible restriction at these stages (review by Illmensee, 1976).

In a different kind of experiment it was shown that whole cells at the cellular blastoderm stage are determined to some extent (Chan & Gehring, 1971). Anterior or posterior blastoderms of one genotype were mixed with whole blastoderms of another. The tissue was then cultured for a while *in vivo* (in the abdomens of adult flies) and then caused to undergo metamorphosis (by implantation into third instar larvae). This kind of procedure is standard in research work on imaginal discs, and therefore it is the disc rudiments or their precursors in the blastoderm which are under investigation. It was found that although structures from all discs were recovered, those marked with the donor genotype were only found either in anterior or in posterior structures, according to the origin of the donor tissue.

Since then, transfers of single blastoderm cells to different regions of host blastoderms have also shown autonomous development. It is not known how precise this determination is, whether for particular segments or for larger subdivisions of the body.

The only cytoplasmic region about which evidence has been obtained by transplantation experiments is the pole plasm. This may not be typical of regional decisions which may occur elsewhere in the cytoplasm since it seems to be laid down as a determinant in the oocyte. Many insects have a posterior region of granular cytoplasm in the egg and cells may form precociously in this region, as soon as nuclei reach it. It has been known since the early years of this century that localised destruction of the pole plasm, or the pole cells, yields sterile adults which have gonads but no germ cells. Also, there exists a mutant of *Drosophila subobscura* called *grandchildless*. This is a maternal-effect mutation which means that a fly expresses the phenotype appropriate to the genotype of its mother rather than appropriate to its own genotype. In this case the affected mothers have sterile offspring (hence *grandchildless*). Maternal-effect mutations are usually assumed to work by producing defects in the egg during oogenesis which cannot be rectified by the paternal genotype after fertilisation. In the eggs of *grandchildless* flies the pole plasm is present but there is a defect of energid migration such that the nuclei do not reach it in time (Fielding, 1967).

Direct evidence that the pole plasm can program the nuclei to become prospective germ cells has been obtained from the experiments of Illmensee & Mahowald (1974) (see Fig. 4.8). Pole plasm was transplanted to the anterior end of host eggs at cleavage stage. Then the cells which formed in the region of the graft were grafted to the posterior end of a genetically distinguishable second host at cellular blastoderm stage. The second host was grown to an adult and then crossed with flies carrying the same recessive marker genotype as the first host. The production of double-recessive progeny proved the presence of functional, marked, germ cells in the chimaeric fly. This experiment has been repeated successfully using pole plasm from unfertilised eggs and late-stage oocytes, suggesting that the determinant is preformed during oogenesis. This result in fact furnishes the only definite proof in any type of embryo for a cytoplasmic determinant laid down locally during oogenesis. In Chapter 5 we shall encounter evidence for some other determinants but many of them appear to be segregated after fertilisation.

Of course biology is replete with examples of similar processes which are carried out in different ways. There are some insect eggs in which destruction of the pole cells does not lead to sterility (Achtelig & Krause, 1971) and there are others in which an entire new germ band complete with germ cells can regenerate from the serosa after destruction of the original embryo (Cavallin, 1971). Such examples indicate the possibility of epigenetic formation of the germ cells even if this is not the normal method.

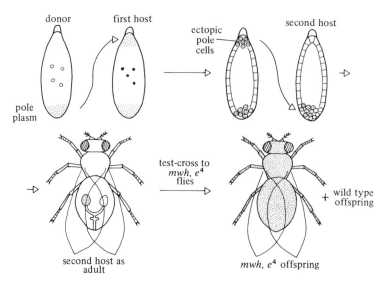

Fig. 4.8. The pole plasm experiment of Illmensee & Mahowald (1974). First, pole plasm is injected into the anterior end of a cleavage stage embryo of a different genotype (*mwh,e*⁴). After cellularisation, the resulting anterior pole cells are transplanted to a posterior position in a second host of different genotype (*y,sn3,mal*). Some of these embryos develop with a chimaeric germ cell population and when test-crossed to *mwh,e*⁴ flies produce double-recessive offspring, proving that some of the nuclei from the first host have become incorporated into the germ line of the second.

Organisation of the longitudinal axis

The gap phenomenon

In several types of insect, experiments have been performed in which the embryo is divided into two by constriction with a hair loop or by pinching with a blunt razor blade. The results follow a common pattern called the 'gap phenomenon'. This is the appearance of a gap in the normal sequence of segments at the constriction (review by Sander, 1976). There is no visible cell death in the region and it therefore seems as though the fate maps of both anterior and posterior fragments are shifted towards the constriction, although there are those who believe that the effect is partly due to damage (Van der Meer, 1979). The gap is large at early stages and becomes smaller at later stages, eventually disappearing altogether. Different insect species show the gap phenomenon to different extents, and in *Drosophila* there has been some disagreement about whether it exists at all. It is shown particularly clearly by the experiments of Jung (1966) on the bean weevil *Bruchidius*. In this type of experiment, as with others discussed in this section, position along the axis of the egg is

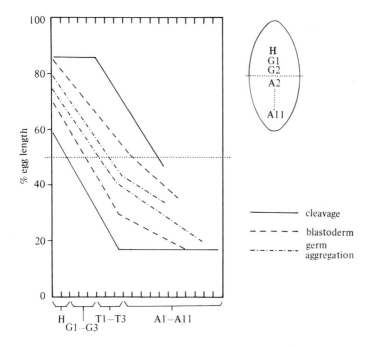

Fig. 4.9. The gap phenomenon. Embryos of the weevil *Bruchidius* were ligated at different levels along the egg axis. For each stage the upper line shows at which structure a posterior fragment will terminate, and the lower line shows at which structure an anterior fragment will terminate. The gap in the sequence of structures is smaller for late ligations than for early ones. An embryo ligated at 50% egg length at blastoderm stage would accordingly lack G3, T1, T2, T3, A1 and so consist of H, G1, G2, A2–11 (inset) where G represents gnathocephalon, T thorax and A abdomen. (After Jung, 1966.)

measured as percentage egg length (EL), with the posterior end as 0% EL and the anterior end as 100% EL. Fig. 4.9 is a graph representing the results. For each stage the upper line shows at which segment the embryo is truncated in the posterior fragment and the lower line shows at which segment the embryo is truncated in the anterior fragment following constriction at a given level. The gap in the sequence is large for early constrictions and has nearly vanished by the 'pre-germ anlage' (or 40 hour) stage.

At the earlier stages of development in *Bruchidius* and several other species, most ligations yield developing embryonic parts only on one side of the constriction (see continuous lines in Fig. 4.9). Since it is usually not possible to get structures on both sides of a constriction the idea has grown up that both of the extreme regions, anterior and posterior, are required for the normal formation of the longitudinal pattern, and in fact the data from the gap phenomenon are fitted quite well by the 'localised source–dispersed sink' gradient model described in Chapter 9. However, Vogel (1978) has shown that it is possible to obtain development in isolated central regions of the egg of the leafhopper *Euscelis*, and he prefers to explain the results on the basis of short-range interactions between several differently specified regions. It is probable that the insects such as *Drosophila* which show little or no gap phenomenon are rather precocious in their specification of regions and develop so rapidly that there is no time for any readjustments.

A posterior signalling centre

Seidel (1929) showed that cautery of the posterior 12% EL region at early stages of development of *Platycnemis* prevented any development of the germ band despite the fact that the prospective region for the germ band lies mainly anterior to this. The posterior pole region became known as the 'activation centre' and its role is shown clearly in more recent experiments on *Euscelis* (Sander, 1960, 1976). The eggs of this species contain a ball of symbiotic bacteria near the posterior pole and this can serve as a marker of the position of the posterior cytoplasm. Sander has combined displacement of the symbiont ball with constriction; the experiments are shown in Fig. 4.10. *Euscelis* shows a gap phenomenon, so an anterior 60% EL fragment isolated during cleavage will typically produce just a procephalon. But when this fragment includes the material marked by the symbiont ball it will form more structures than this, often complete germ bands. The posterior 60% EL fragment will usually produce a sequence of structures from gnathocephalon to abdomen, but when it includes the material marked by the symbiont ball at its anterior end it produces an inverted or a mirror-duplicated set of posterior structures. If the symbiont ball is moved forwards during cleavage and the embryo is constricted some hours later, at the early germ anlage stage, then both effects are produced no matter which side of the constriction the symbiont ball lies. This implies that its influence can spread in the intervening period. Sander has argued that these results suggest that the posterior region is the source of a graded signal, and that together with other evidence they support his double-gradient model for longitudinal specification (Sander, 1976).

Although such manipulations are not possible in *Drosophila*, there is evidence for a longitudinal gradient which controls the disposition of the principal body structures from a mutant called *bicaudal* (Nüsslein-Volhard, 1977). *Bicaudal* (*bic*) is a maternal-effect recessive lethal: *bic/bic* females are viable but lay some defective eggs. Although the eggs do not hatch they develop far enough to reveal a most remarkable phenotype, which is a mirror-symmetrical arrangement of abdominal segments in

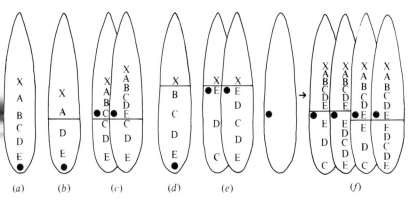

Fig. 4.10. Properties of the posterior pole plasm in the leafhopper *Euscelis*. (*a*) Normal sequence of structures in the germ band: X, serosa; A, procephalon; B, gnathocephalon; C, thorax; D and E, abdomen. (*b*) Following ligation a gap appears in the pattern. (*c*) A more complete set of structures is formed by the anterior fragment if posterior cytoplasm is moved forward before the ligation. (*d*) A different gap results from a more anterior ligation. (*e*) Partial or complete polarity reversal of structures within the posterior fragment can be produced by displacement of the posterior cytoplasm. (*f*) If a time interval is left between the displacement and the ligation then both effects can be produced together. (After Sander, 1960.)

between two posterior ends (Fig. 4.11). The number of segments in between the ends is variable, and the pattern is not always symmetrical, the 'normal' posterior sometimes being larger. These variations can be explained on the assumption that a monotonic gradient present in the normal oocyte becomes altered towards a U shape. The mutation has a variable penetrance which is much affected by environmental conditions and for this reason it is not thought to be a defect in the signalling system itself but a perturbation arising from a defect elsewhere. Another maternal-effect mutant, which has the opposite effect, has recently been described by Lohs-Schardin (1982). This is called *dicephalic* and homozygous females lay occasional eggs of double anterior structure. It is thought that these arise from abnormal oocytes which have nurse cells at both ends instead of just at the anterior end.

Embryos with double abdomens can also be produced by experimental manipulation of the eggs of the midge *Smittia* (Kalthoff & Sander, 1968; Kalthoff, 1971, 1973; Fig. 4.12). They can result from irradiation of the anterior end with UV light, from centrifugation, or from puncture of the anterior end. Since a variety of treatments is effective, Kalthoff feels that double-abdomen formation arises from the destruction of something which he calls an 'anterior determinant'. The action spectrum for UV inactivation has been measured and shows a peak at 280 nm, suggesting a

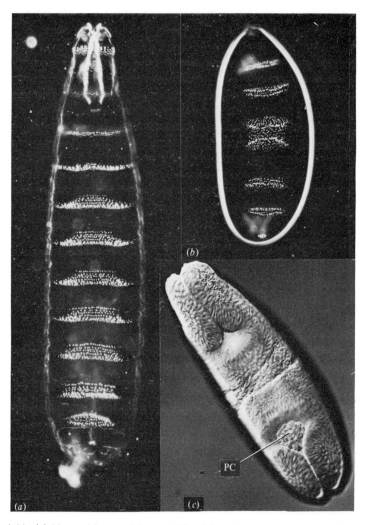

Fig. 4.11. (*a*) Normal larva of *Drosophila*. (*b*) Embryo produced by a *bicaudal* female, consisting of a mirror symmetrical duplication of the most posterior segments. (*c*) Nomarski photomicrograph of *bicaudal* gastrula: the pole cells (PC) are not duplicated. (Photographs kindly provided by Dr C. Nüsslein-Volhard.)

protein target. On the other hand the UV effect can be reversed by exposing the eggs to visible light and the characteristics of this photo-reversal suggest an RNA target. The regime of centrifugation which produces double abdomens in fact stratifies the eggs into three zones of different density: lipid, cytoplasm and yolk granules. According to the conditions, double cephalons and inverted embryos can be produced as well as double abdomens (Yajima, 1960; Rau & Kalthoff, 1980). So gross

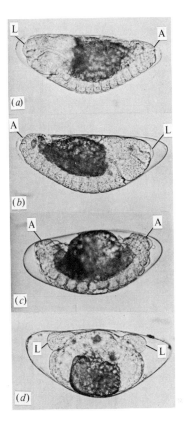

Fig. 4.12. Normal and abnormal body patterns of *Smittia* embryos developing from centrifuged eggs. (*a*) Normal embryo, (*b*) inverted embryo, (*c*) double abdomen, (*d*) double cephalon. A, anal papillae; L, labrum (most anterior head structure). The anterior end of the egg shell containing the micropyle is oriented to the left, the dorsal side is up. All four patterns developed during cleavage with the posterior pole pointed in the direction of centrifugal force. (From Rau & Kalthoff (1980); photographs kindly supplied by Dr Kalthoff.)

disruption of the cytoarchitecture can clearly shift around the 'reference points' for the signals controlling longitudinal pattern. It is worth noting that the double abdomens in *Smittia* do not have two sets of pole cells. These are present only in the original abdomen and this is further evidence for the independence of the pole plasm mechanism from other pattern-forming processes.

The bithorax *system in* Drosophila

The data discussed in the previous sections have indicated the existence of one or more graded signals controlling the disposition of body parts in

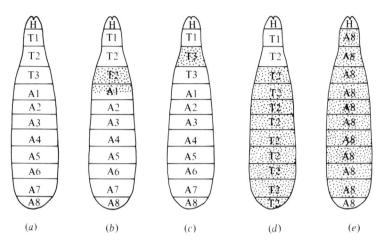

Fig. 4.13. Phenotypes of *Drosophila* larvae carrying pattern mutations. (*a*) Wild type, (*b*) *Ultrabithorax*, (*c*) *Contrabithorax* (effect not visible until adult stage), (*d*) P9, (*e*) *Polycomb*. H, head; T1–T3, thoracic segments; A1–A8, abdominal segments.

the longitudinal axis, and the actual mechanism of this sort of process will be considered in Chapter 9. But a gradient can only produce a pattern if the cells are competent to respond in different ways at different levels. These responses can be defective and give rise to pattern mutants in which structures are formed in inappropriate positions. The best-known set of such mutations is the *bithorax* series in *Drosophila melanogaster*. It has been known for over 50 years that they affect adult flies, but only recently has their mode of action begun to become clear through a study of their effect on the larval anatomy (Lewis, 1963, 1978). The larva of *Drosophila* is shown diagrammatically in Fig. 4.13. It has a head, three thoracic segments, and eight abdominal segments. There are no legs, but thoracic and abdominal segments can be distinguished by differently shaped ventral setal bands, by the presence of Keilen's organs on the thoracic segments, and spiracles in mesothorax and eighth abdominal segment.

The *bithorax* complex (*BXC*) is a series of genes on the right arm of chromosome III and the genetic map indicates that the order of loci on the chromosome corresponds roughly with the order of body segments affected in the various mutants. Most of the mutants are now known to result from the insertion into the gene of an extra segment of DNA. Since the genetic data are highly complex the following account is inevitably somewhat oversimplified.

If the whole *BXC* is deleted from both chromosomes (P9 deletion) the embryos die before hatching but can be seen to consist of a head, prothorax and then ten copies of the mesothorax; this is consistent with the proposition that the products of *BXC* are required to code for all

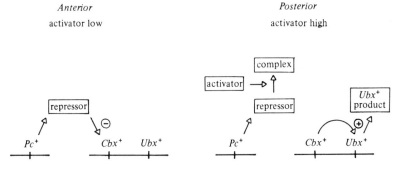

Fig. 4.14. Possible mechanism of action of the *bithorax* complex. In the anterior part of the embryo the activator is at an insufficient level to derepress *Ubx*⁺. In the posterior part it does derepress *Ubx*⁺, and the derepression will normally occur in the metathorax and all segments posterior to it.

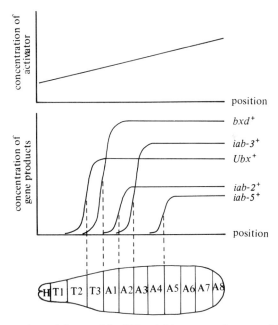

Fig. 4.15. Extension of the model of Fig. 4.14 to several genes of the *bithorax* complex. The levels of the various gene products in regions of activation are chosen arbitrarily.

segments more posterior than the mesothorax. The operation of the system is best understood by considering the formation of the metathorax (Figs. 4.14, 4.15). A mutant *Ubx* (*Ultrabithorax*) in homozygous form will convert the larval metathorax and part of the first abdominal segment to mesothorax. Such animals die as larvae or pupae. So the normal gene, Ubx^+, presumably codes for the metathorax and is normally switched off in the mesothorax and on in the metathorax. If a short chromosome segment containing Ubx^+ is combined with a homozygous deletion of the whole *BXC* then the embryos are formed with head, prothorax, mesothorax and nine copies of the metathorax. So this suggests that Ubx^+ is normally switched on in all segments posterior to the mesothorax.

A mutation *Cbx* (*Contrabithorax*) has the reverse effect to *Ubx*, converting the mesothorax to a second metathorax. This mutation only shows its effect if it is on the same chromosome as a wild type Ubx^+ gene. This is the 'cis' configuration: $Cbx,Ubx^+/Cbx^+,Ubx$. In the opposite 'trans' configuration – $Cbx,Ubx/Cbx^+,Ubx^+$ – there is no effect. Cbx^+ is therefore thought to be a regulatory gene which normally represses Ubx^+ in the mesothorax. For those readers familiar with the *lac* operon of *E. coli*, Cbx^+ is analogous to the operator, *Cbx* is an operator-constitutive mutant, and Ubx^+ is the structural gene controlled by the operator. The repressor is supposedly the product of a gene outside the *BXC*, one candidate for which is the gene whose mutants are called *Polycomb* (*Pc*). *Polycomb* homozygotes have most of their segments transformed towards eighth abdominal and die as late embryos, and this extreme posteriorisation is what would be expected if the entire *BXC* were derepressed. Similar properties have recently been shown for the mutant *extra sex combs* (Struhl, 1981). The inducer is the hypothetical morphogen which forms a monotonic gradient along the egg with the high point at the posterior end. So, in normal development Ubx^+ will be repressed anteriorly and derepressed posteriorly, the threshold of activation presumably coinciding with the normal boundary between the second and third thoracic segments.

There are at least four other recessive mutations homologous to *Ubx*. Each converts a particular segment i into a second copy of $i - 1$. Each is apparently controlled by an operator-like gene whose mutations are dominant and convert segment $i - 1$ into a second copy of i. Assuming that they are similar to Ubx^+, the situation in normal development is shown in Fig. 4.15. The mutations that have been found to date are listed in Table 4.4. There are also some other mutations in the complex which specifically affect the imaginal structures.

A pseudoallelic locus which in some respects resembles the *BXC* is the *E* locus found in the silkworm *Bombyx mori* (Tazima, 1964). In particular there is a mutation E^N which in one dose produces a partial anteriorisation of abdominal segments. In double dose it is a late-embryo lethal and converts all the abdominal segments to a thoracic morphology. This bears an obvious resemblance to the deletion P9 in *Drosophila*. However, the other mutations of the *E* complex are not so similar to those of the *BXC*. There are some which produce anterior and some posterior displacements

Table 4.4 *Mutations of the* bithorax *complex in* Drosophila

Name	Symbol	Action	Presumed function of wild type gene
Dominant mutations			
Contrabithorax	*Cbx*	MS → MT	Control of Ubx^+, bxd^+
Hyperabdominal	*Hab*	MT, A1 → A2	Control of $iab\text{-}2^+$
Ultra-abdominal	*Uab*	A1, A2 → A3	Control of $iab\text{-}3^+$
Miscadestral pigmentation	*Mcp*	A4 → A5	Control of $iab\text{-}5^+$
Recessive mutations			
Ultrabithorax	*Ubx*	MT, A1 → MS	Codes for MT
bithoraxoid	*bxd*	A1 → MT	Codes for A1
infra-abdominal 2	*iab-2*	A2 → A1	Codes for A2
infra-abdominal 3	*iab-3*	A3 → A2	Codes for A3
infra-abdominal 5	*iab-5*	A5 → A4	Codes for A5

MS, mesothorax; MT, metathorax; A, abdominal.

of various cuticular structures but they are all dominant. In the *BXC* the rule is that posteriorisation is dominant and anteriorisation recessive; and the functional genes are believed to be switched on in a sequence from anterior to posterior.

Segmentation

The *bithorax* mutations change the character of segments but do not change their total number. This is particularly clearly shown in the deletion P9 which lacks the whole *BXC* and develops ten copies of the mesothorax, and it suggests that some other system is responsible for establishing the repeating pattern that we visualise as segmentation.

As we have seen above there is some evidence from clonal analysis that the segment boundaries become determined during the blastoderm stage in the milkweed bug *Oncopeltus*. In *Drosophila* no stage is known at which X-ray-induced clones visible in the adult cuticle will cross segment boundaries, but this is probably because embryos are not viable if they are irradiated before the blastoderm stage. Curiously enough, clones in the central nervous system will cross segment boundaries when induced as late as 10 hours after fertilisation, which in *Drosophila* is well after the visible segmentation of the larval hypoderm (Ferrus & Kankel, 1981).

Recently a series of mutants of *Drosophila* have been described which are defective in normal segmentation (Nüsslein-Volhard & Wieschaus, 1980). All the mutations are recessive and in double dose are lethal: the animals survive, however, until late embryo or early larval stage, by which time the pattern of segmentation is clearly indicated by the arrangement of ventral setal bands (Fig. 4.16). Perhaps rather surprisingly they are all zygotic in action so it is the genotype of the embryo itself which controls the pattern rather than that of the mother (there is in addition a maternal effect in the case of the mutation *fused*). The mutations fall into three groups: those affecting every repeating unit, those affecting every other repeating unit and those deleting major regions of the body plan. The term 'repeating unit' rather than 'segment' is used because the mutations' area of activity can span the segment boundary. In each case the mutation's effect involves a deletion of a part of the pattern and, if this should result in an apposition of non-equivalent parts of the repeating unit, substitution of a duplicated part. For example, *gooseberry* replaces the posterior half of each segment by a mirror image of the anterior half. *Hedgehog* is similar, but the region duplicated is a little more posterior and so does not include the segment boundary. The embryo therefore has no segment boundaries although the periodicity is still visible because of the setal bands. By contrast, *patch* duplicates a region spanning the segment boundary and thus has twice as many boundaries as usual.

The 'pair rule' mutations do not involve any duplication, presumably because the deletions which they cause join homologous levels of the repeating pattern. *Even-skipped* deletes segments T1, T3, A2, A4, A6 and A8 (where T represents thoracic and A abdominal) and *odd-skipped* deletes the complementary segments H (head), T2, A1, A3, A5 and A7. *Paired* deletes the posterior part of each odd segment and the anterior part of each even segment. Hence each segment becomes a composite of anterior odd and posterior even. These mutations suggest strongly that there is a level of the pattern which is repeated every two segments.

The discovery of two distinct sets of mutants, one in which the character of segments is affected and the other the number of segments, does suggest that different processes are at work forming these two aspects of the body plan. Of course, reasoning from mutants is a tricky business. If we know nothing about a mutant except that it lacks segment *n*, it does not follow by any means that the wild type gene product codes for segment *n*, since the gene may have some quite different function and the pattern abnormality may be a very indirect consequence. Nonetheless, the availability of the mutants provides a possible avenue for the application of modern recombinant DNA techniques to the study of regional specification. We await the results with interest.

The dorsoventral pattern

Everything so far has concerned the long axis of the organism. But there is another dimension to the body plan: the dorsoventral axis. At the

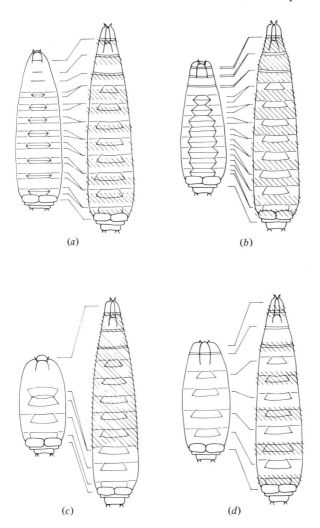

Fig. 4.16. Mutations of *Drosophila* affecting the segmental repeating unit. Each diagram depicts the derivation of the mutant phenotype from the normal larval anatomy. (*a*) *patch*: each segment is replaced by duplication of a region overlapping the boundary so that the total segment number is doubled. (*b*) *gooseberry*: each segment is replaced by a duplication of its own anterior half. (*c*) *kruppel*: most of the pattern is missing and segment A6 is duplicated. (*d*) *even-skipped*: the even-numbered segments are missing. (After Nüsslein-Volhard & Wieschaus, 1980.)

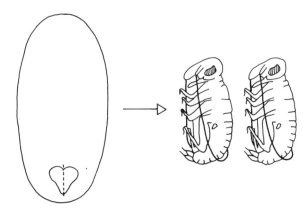

Fig. 4.17. Twins produced by longitudinal constriction of germ anlage stage
embryos of *Tachycines*. (After Krause, 1958.)

blastoderm stage the embryo consists of an ellipsoidal sheet of cells
arranged around the yolk mass. In fate map terms the prospective regions
from ventral to dorsal are: mesoderm, neurogenic region, larval cuticle
(or hypoderm) and extra-embryonic serosa. At some stage a specification
must occur into at least this many regions.

In the investigations of Krause (1958) this stage turned out to be very
late. He cut the germ anlage of the camel cricket *Tachycines* into
longitudinal halves and was able to produce complete twins up to the
gastrula stage (Fig. 4.17). On the basis of the orientation of the twins he
distinguished an early regulative phase in which each half of a germ anlage
would undergo gastrulation in its own midline, and a late phase in which
a half gastrula 'induced' the other half from the amnion. At still later
stages it was possible to obtain duplications in the abdomen, presumably
because of its late formation from a growth zone.

Tachycines is of course a very slowly developing insect, although
similar results have been obtained by cutting germ anlage of *Bombyx
mori* (Krause & Krause, 1965) and a number of species are able to recover
from lateral irradiation at early stages. As in other embryos the rule seems
to be that in slowly developing species the determinative events occur
later relative to morphological stage than in rapidly developing species.

Drosophila is, accordingly, considered to be rigidly mosaic with respect
to cutting and irradiation experiments. There are, however, some mutants
which seem to represent defects in the mechanism responsible for
dorsoventral organisation. One such is called *dorsal* and is a maternal
effect mutant (Nüsslein-Volhard *et al.*, 1980). One copy of the gene
generates embryos lacking mesoderm and ventral hypoderm. Two copies
generate featureless tubes of hypoderm lacking all dorsoventral polarity
(Fig. 4.18). Defects produced by laser microbeam irradiation show that the

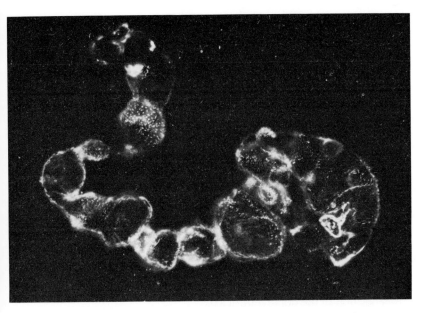

Fig. 4.18. Embryo produced by homozygous *dorsal* female. All ventral structures are absent. (Photograph kindly provided by Dr C. Nüsslein-Volhard.)

fate maps of these mutants are drastically altered, with a loss of ventral structures. Since there is apparently no regional cell death the authors argue that a dorsoventral gradient has been disrupted, a gradient similar to but independent of the gradient along the longitudinal axis.

Comparison with amphibian embryos

The observable course of embryonic development is quite different in insects and amphibians and so, of course, is the anatomy of the larval and adult animals themselves. There are, however, certain general similarities which are unrelated to their particular anatomy. In both cases some disruptions of the egg cytoarchitecture can produce mirror-symmetrical duplications of the embryonic pattern, which suggests that the earliest developmental decisions may involve cytoplasmic determinants. In both cases some regions of the embryo appear to be specified by the cleavage stages, as judged by isolation experiments, but the nuclei remain totipotent and capable of directing the development of an enucleated egg into which they are transplanted. This again points to the cytoplasm as the initial location of developmental commitment.

In amphibians from fertilisation and in insects from the syncytial blastoderm stage the fate maps indicate that cell clumps remain coherent despite the considerable morphogenetic movements involved in gastrulation. This shows that the general body plan must emerge from some process of regional specification rather than by random cell differentiation followed by sorting out.

Regional specification after the earliest stages clearly involves inductive interactions and two important organising centres are represented by the dorsovegetal region of the amphibian blastula (the organiser) and the posterior region of the insect cleavage syncytium (the activation centre). Both are able to induce mirror-symmetrical duplications after transplantation, providing evidence for the gradient types of model for embryonic induction (see Chapter 9). At first sight a gradient mechanism might seem unlikely for those insects in which much of the body is formed by a posterior growth zone. However, in Chapter 10 it will be shown that at the biochemical level the mechanisms of a 'gradient' and a 'progress zone' may actually be rather similar.

Twinning is possible in some species of amphibians and insects following median division but not following transverse division, indicating that in both cases the determination of parts in the anteroposterior axis precedes that in the dorsoventral axis.

The extensive range of pattern mutants which exists among the insects does not appear to have a counterpart among the amphibians or any other vertebrate group. It is normally thought that the reasons for this are technical rather than fundamental. The amphibian life cycle is too slow for genetic research and in a small mammal such as the mouse any pattern mutants as severe as those discussed above would probably be aborted at a very early stage and so never be detected by the experimenter. The availability of the pattern mutants among the insects offers the possibility of applying the techniques of recombinant DNA research to early embryonic development and characterising the determined states in terms of the activity of combinations of regulatory genes. At present it is still too early to attempt to reconstruct a genetic program of development from a study of mutant phenotypes because it is clear that the situation is a complex one. A lot more characterisation needs to be done of the existing mutants, and it is probable that many more mutants remain to be discovered.

General references

Counce, S. J. (1972). The causal analysis of insect embryogenesis. In *Developmental System: Insects*, vol. 2, ed. S. J. Counce & C. H. Waddington, pp. 1–156. Academic Press, London.

Sander, K. (1976). Specification of the basic body pattern in insect embryogenesis. *Adv. Insect Physiol.* 12, 125–238.

Lawrence, P. A. (ed.) (1976). *Insect Development. Royal Entomological Society Symposium 8*. Blackwell Scientific, Oxford.

5

Development with a small cell number

In this chapter we shall examine the experimental embryology of several diverse types of animal: molluscs, annelids, ascidians, ctenophores and sea urchins. With the exception of sea urchins, which are included more as a matter of convenience, these groups can be collectively contrasted to the embryos considered previously in so far as the key decisions of early development seem to be made at a very early stage when there are only a few cells in the embryo. This may mean that each individual cell has a unique identity in terms of its biochemical properties and corresponds to a zone of tens or hundreds of similarly committed cells in a vertebrate, insect, or sea urchin embryo.

The small cell number means that all embryos of a given species are identical and makes it possible in principle to construct fate maps of very high precision by direct observation of the cell lineage. A number of studies of this type were carried out around the turn of the century and some, such as Conklin's study of the ascidian *Styela* (Conklin, 1905*a*), are masterly works still referred to today. More recently a major effort has been made to obtain a complete description of the cell lineage of the nematode *Caenorhabditis elegans* (Sulston & Horvitz, 1977; Deppe *et al.*, 1978). This organism was especially chosen for detailed study by molecular biologists because of its suitability for genetic analysis. Unfortunately it is not very suitable for micromanipulation and so not much has yet been possible in the way of experimental embryology. The interactions which have been discovered (Kimble, 1981) occur in larval development and are therefore outside the scope of this book. However, the descriptive study has resulted in a fate map which traces the lineage of every somatic cell from the egg to the adult. It would certainly be desirable if a fate map of this quality were available for other types, such as molluscs, for which there already exists a good body of evidence about the mechanism of early developmental decisions.

The case for using these animal species as experimental models is partly that they are felt to be 'simple' in some sense, but mainly that they offer favourable material for the study of cytoplasmic localisation. The role of unequal cell divisions in vertebrate development is not really clear but they are often thought to be important for the reproduction of those cell populations which undergo continuous turnover in the adult, such as skin, intestinal epithelium or blood. It is normally assumed that the cytoplasm of the stem cell must become partitioned into two zones before cell division and that in one daughter the cytoplasm maintains the nucleus as a stem cell nucleus while in the other it causes the nucleus to set off down a pathway involving a finite number of further divisions followed by terminal differentiation. Although embryos are small, they are a lot larger and more accessible than stem cells in the adult and it is to be hoped that they can tell us something about the association of cell division and the segregation of developmental potency.

Molluscs

Normal development and fate map

Much recent work has been performed on the marine mud snail, *Nassarius obsoletus*, usually known by its subgeneric name of *Ilyanassa*, and it is the development of this species which is described here.

The egg is about 170 μm in diameter and develops quite slowly, taking about 7 days to reach the veliger larva stage at which the general molluscan body plan is first visibly differentiated. Fertilisation occurs before oocyte maturation and at the first and second meiotic divisions a cytoplasmic protrusion called the polar lobe appears at the vegetal end of the egg; it is not to be confused with the polar bodies, which are as usual extruded near the animal pole. The polar lobe reappears at the first and second cleavages and is a particularly prominent feature of early development. The first cleavage is unequal and gives rise to one cell called AB and a somewhat larger one called CD (Fig. 5.1). During the first cleavage the lobe attachment becomes highly constricted and since the lobe is as large as both the forming blastomeres the embryo appears superficially to consist of three cells; this stage is accordingly called the 'trefoil'. The lobe is absorbed into the CD blastomere and accounts for its greater size. The lobe appears again at the second cleavage, in which the four blastomeres A, B, C and D are formed, and passes into the D blastomere which thereby becomes larger than the other three.

The plane of cleavage is altered to the equatorial for the third division. Each of the four blastomeres A, B, C and D, now called macromeres, cuts off a small cell called a micromere at its animal end. Once this has happened the macromeres are relabelled 1A, 1B, 1C, 1D and the micromeres are called, respectively, 1a, 1b, 1c and 1d. This type of division is repeated several times, each ring of micromeres being called a quartet. So the second quartet comprise 2a, 2b, 2c and 2d and the macromeres then become 2A, 2B, 2C and 2D. The cleavage pattern is called spiral because each quartet of micromeres is somewhat rotated relative to the macromeres. When looked at from the animal pole the first quartet is formed by right-handed or dexiotropic cleavages, the second by left-handed or laevotropic cleavages, and the orientation of cleavages continues to alternate for subsequent quartets.

Each quartet of micromeres also divides at a time somewhat after the corresponding macromere division. The offspring are designated by numbers after the letter, sometimes written as superscripts; so, for example, 1a becomes 1a1 and 1a2, and 2c becomes 2c1 and 2c2. The same nomenclature is followed for subsequent divisions, so 1a1 becomes 1a11 and 1a12 and 2c2 becomes 2c21 and 2c22. By the time the first quartet has become 16 cells the eight cells at the animal pole (1a11, 1a12, 1b11, 1b12, 1c11, 1c12, 1d11, 1d12) are arranged in a characteristic pattern called the molluscan cross. The complete cell lineage for these early stages is shown in Fig. 5.2.

Particular attention should be paid to the micromere 2d which is called the somatoblast and forms much of the veliger body, particularly the shell gland. Even more noteworthy is the mesentoblast 4d. This arises somewhat before the other cells of the fourth quartet, and is much larger than they are. After formation it divides in the plane of bilateral symmetry of the embryo to form two mesentoblasts, one on the right and one on the left side (Fig. 5.1). Each of these then follows the cell lineage depicted in Fig. 5.3. The enteroblasts contribute to the gut, and the mesoblasts, particularly the stem cells called the mesoblastic teloblasts, give rise to internal mesodermal bands which subsequently form the larval muscles and mesenchyme. Some mesoderm is, however, also formed from the micromeres of the second and third quartets.

Gastrulation occurs by a spreading of the micromere cap to cover the whole embryo. After this the embryo becomes a trochophore which has a ciliated apical end derived from the animal micromeres and ciliated bands around the equator. In later development the remnants of the blastopore become shifted anteriorly and a stomodaeal invagination arises at its site to become the mouth. The dorsal ectoderm forms the shell gland and the ventral ectoderm the foot. The stomach and digestive gland are formed from the endodermal cells and the intestine from the mesentoblast. The preoral region, derived from first and second micromere quartets, forms the bilobed locomotary organ called the velum which bears paired eyes dorsally and paired statocysts ventrally. The veliger larva thus represents the definitive body plan of the organism. It hatches after 7–8 days and in later development the bilateral asymmetry becomes more pronounced and the shell assumes its characteristic right-handed spiral form.

The fate maps of molluscs are largely based on descriptive studies of cell lineage which cannot be carried beyond the trochophore stage. It is to be hoped that more accurate fate maps for veliger larva structures will

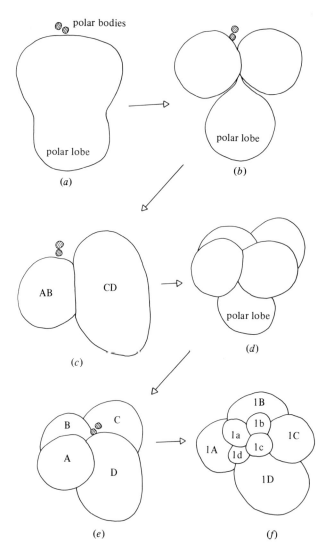

Fig. 5.1(*a*)–(*f*). For legend see opposite.

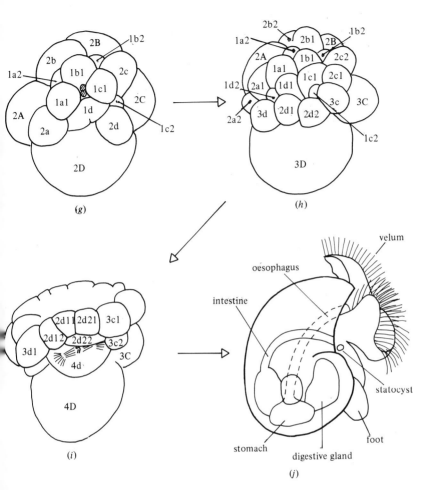

Fig. 5.1. Normal development of *Ilyanassa* from first cleavage to veliger larva. (*a*) Fertilised egg, (*b*) trefoil stage, (*c*) 2-cell stage, (*d*) second cleavage, (*e*) 4-cell stage, (*f*) first micromere quartet formed by dexiotropic cleavage, (*g*) and (*h*) continuation of spiral cleavage, (*i*) median division of 4d to give two mesentoblasts, (*j*) veliger larva. (After Clement, 1952.)

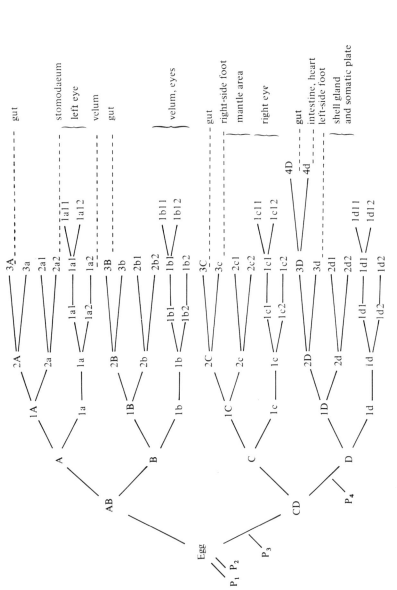

Fig. 5.2. Cell lineage of *Ilyanassa* up to the 29-cell stage. P represents the polar lobe which appears four times

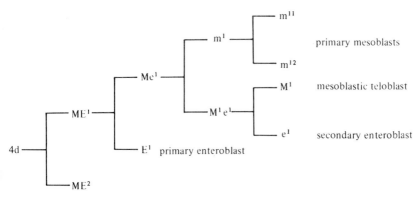

Fig. 5.3. Cell lineage of the mesentoblast.

soon be constructed with the aid of the high molecular weight intra-cellular tracers which are now available. The assignments made on Fig. 5.2 are drawn from four sources: Conklin's description of the cell lineage of *Crepidula* (Conklin, 1897), Van Dongen & Geilenkirchen's (1974) description of *Dentalium*, defect experiments by Clement (Clement, 1963, 1967) and marking experiments using carbon particles (Cather, 1967).

The description given for *Ilyanassa* is broadly applicable to other molluscs used in experimental embryology except for cephalopods. For species mentioned in this chapter it should be noted that the scaphopod *Dentalium* follows a very similar course of early development. The other species are, like *Ilyanassa*, gastropods. *Crepidula* and *Bithynia* differ in that their polar lobes are much smaller, while *Limnaea* and *Patella* have no polar lobe at all. In the latter forms the early cleavages are symmetrical and the D blastomere can only be named retrospectively once the mesentoblast has been formed. The description of the early stages is also applicable to some annelid worms, which undergo a similar spiral cleav-age and gastrulation as far as the trochophore larva.

Role of the egg cytoplasm

Many experiments on small invertebrate embryos have involved the use of actinomycin D. This is an inhibitor of RNA synthesis and the idea is that events which occur in the presence of actinomycin do not require new transcription of genes but are carried out by components formed before the treatment began. It has been shown that *Ilyanassa* will develop normally up to gastrulation in the presence of concentrations of actinomy-cin which block essentially all RNA synthesis (Newrock & Raff, 1975). Protein synthesis in *Ilyanassa* has recently been examined by Collier &

McCarthy (1981) using two-dimensional gels. They found that out of 98 proteins examined, 12 were formed only by the cleavage stages and 24 only by the mesentoblast stage. The protein synthesis pattern of cleavage-stage embryos was not qualitatively affected by actinomycin but a number of differences were apparent in embryos reared in actinomycin up to the mesentoblast stage. Despite this, 18 of the 24 proteins specific to the mesentoblast stage were still synthesised. This implies not only that the protein synthesis in the egg is directed by maternal mRNA but that protein species which appear only at later stages may be formed from maternal mRNA as well. It must be remembered, however, that actinomycin D experiments have been criticised on various grounds; for example it is impossible to inhibit absolutely 100% of RNA synthesis, and there are always some toxic side-effects to contend with.

Molluscan eggs have also been extensively studied by electron micro-scopy. Ultrastructural studies on the polar cytoplasm have revealed different features in different species (Dohmen & Verdonk, 1979). The lobe of *Ilyanassa* includes some unusual double-membrane vesicles con-taining dense material. The lobe of *Dentalium* contains 'multisheet vesicles'. The lobe of *Crepidula* has many folds on the plasma membrane and that of *Bithynia* contains an RNA-rich 'vegetal body' which appears during oogenesis (Fig. 5.4).

An important feature of the early development which is clearly under maternal control is the orientation of the spiral cleavage. Normally the third cleavage is dexiotropic, the fourth laevotropic, the fifth dexiotropic, and so on. Nearly all adult snail shells are right-handed spirals when viewed from the tip. However, occasionally an individual or local race of left-handed snails may be found and their embryos undergo a laevotropic third cleavage, dexiotropic fourth, and so on. Breeding experiments using right- and left-handed variants of the freshwater snail *Limnaea* have shown that the handedness of the shell is determined by a single gene and that right-handed is dominant over left-handed. However, the handed-ness of an individual is not determined by its own genotype but by that of the mother (Boycott & Diver, 1923; Sturtevant, 1923). As we have discussed in Chapter 4, a maternal effect implies that the feature under consideration is laid down before fertilisation, presumably during the development of the oocyte. Since the left-handed symmetry is apparent as early as the third cleavage this seems quite reasonable. It has recently been shown that left-handed embryos can be made right-handed by microinjection of cytoplasm from a right-handed embryo. As expected from the genetics, the reverse procedure has no effect (Freeman & Lundelius, 1982). Presumably the left-handed embryos have some defect in a component of the cytoskeleton which causes the spindles to tilt in the opposite direction from usual, although the molecular basis remains undiscovered.

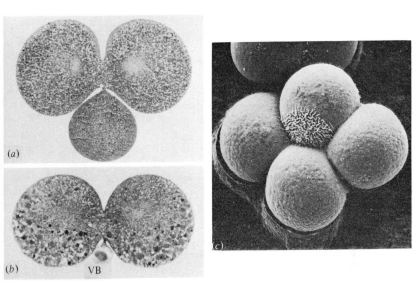

Fig. 5.4. Polar lobes of three molluscan species. (*a*) Section of trefoil stage of *Dentalium*. (*b*) Section of trefoil stage of *Bithynia*. VB, vegetal body. (*c*) Scanning electron micrograph of 4-cell stage of *Crepidula*. The former lobe is visible as a puckered region on the D blastomere. (Photographs kindly provided by Dr M. R. Dohmen.)

Regional organisation

Given that most of the early protein synthesis takes place on maternal templates, that there is some inhomogeneity of ultrastructure, and that at least one aspect of pattern (the orientation of spiral cleavage) is determined during oogenesis, it seems reasonable to ask to what extent the fertilised egg, or even the oocyte, is already partitioned into different regions. This would presumably be accomplished by the localisation of cytoplasmic 'determinants', each of which would regulate the activity of the nuclei which came within their influence in such a way as to produce the correct arrangement of cell types in the normal early embryo. The answer to this question has been sought for mollusc embryos and for the other types considered in this chapter by four main kinds of experiment: (1) redistribution of egg cytoplasm by low-speed centrifugation; (2) removal of parts of the egg cytoplasm; (3) removal of blastomeres or portions of blastomeres and study of the development of the remainder of the embryo, and (4) study of the development of isolated blastomeres.

Low-speed centrifugation can stratify the egg cytoplasm to give bands of lipid droplets, soluble cytoplasm and yolk granules, arranged perpendicular to the centrifugal direction. This treatment has been shown to have no effect on the subsequent development of fertilised eggs of several

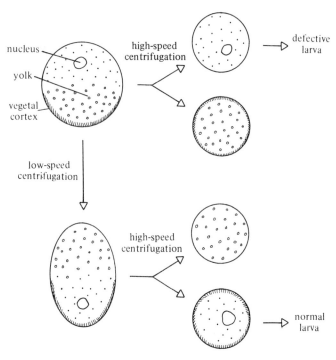

Fig. 5.5. The centrifugation experiment of Clement (1968), which suggests the necessity of vegetal cortex material for formation of a normal larva.

molluscs including *Crepidula* and *Cumingia* (Morgan, 1927) and *Dentalium* (Verdonk, 1968). This implies either that there is no regionalisation of cytoplasm before the first cleavage or that if there is then the factors responsible are not affected by centrifugation. It is often thought that the 'cortex', or cytoplasm just below the plasma membrane, is not redistributed and could therefore be the site of regionalisation (Raven, 1966).

When an unfertilised egg of *Dentalium* is cut in two then the vegetal half, after fertilisation, is capable of developing into trochophores of normal proportions but reduced size (Wilson, 1904*a*). This shows two things: first that the animal cytoplasm is not essential for formation of the normal pattern, and secondly that the sizes of parts can be scaled down in proportion to the size of the entire embryo.

Clement (1968) was able to separate the properties of the cortex and the mobile cytoplasm by experiments using two types of centrifugation (Fig. 5.5). Here and elsewhere in the present chapter we shall take 'low-speed' centrifugation to be a speed sufficient to stratify the egg cytoplasm while 'high-speed' centrifugation causes the egg to break into two parts. Animal halves containing the nucleus were isolated by high-speed centrifugation in 30% raffinose. In this solution they remain in suspension while the yolk-rich vegetal halves form a pellet. Nucleated vegetal halves were isolated by a two-step procedure. First the eggs were stratified by embedding them in gelatin and centrifuging at low speed towards the animal pole. This makes the yolk enter the animal half and the nucleus and soluble cytoplasm enter the vegetal half. Then they were centrifuged at high speed in raffinose, which causes them to break in two and leave suspended the vegetal cortical regions containing the nuclei. When allowed to develop the vegetal halves could form normal or nearly normal veligers of reduced size, while the animal halves formed defective larvae similar to those produced by removal of the polar lobe at the trefoil stage (see below for discussion of this).

Guerrier (1970*a*, *c*) carried out experiments in which the plane of the first cleavage was altered by compression. In forms which possess a polar lobe he found that a cleavage which distributed some lobe material to each blastomere gave rise to veligers with many structures duplicated. On the other hand in forms with equal cleavage and no polar lobe the plane of the first cleavage was without effect on later development.

Taken together these results suggest that, at least in the lobe-forming species, the morphogenetic properties of the D blastomere lineage may be specified by some localisation in the vegetal cortex of the fertilised egg. By contrast the amount and composition of the soluble cytoplasm have little effect on the course of development.

Lobe removal and blastomere isolation

At the trefoil stage the polar lobe is connected to the CD blastomere by a thin stalk and so it is a reasonably simple manipulation to cut it off. This operation was first performed by Crampton (1896) on *Ilyanassa* and the studies have been extended more recently by Clement (1952), Atkinson (1971) and Van Dongen (1976).

Lobeless embryos are quite viable and they continue to develop at the same rate as controls for many days. However, their organisation is greatly disturbed. At the 2-cell stage the CD blastomere is no larger than AB, and at the 4-cell stage D is no larger than A, B and C. Various peculiarities of the D lineage which are apparent in normal embryos are absent in lobeless ones. In particular the cell 4d is no different from 4a,b,c. It does not form earlier, it is no larger, and it does not initiate the sequence of cleavages leading to the formation of the mesodermal bands. The lobeless embryos thus form no mesodermal bands although they do develop some muscle and mesenchyme, presumably from the second and third micromere quartets. They also have no eyes, no foot and usually no shell. They have velar cilia all over the place and the main segments of the

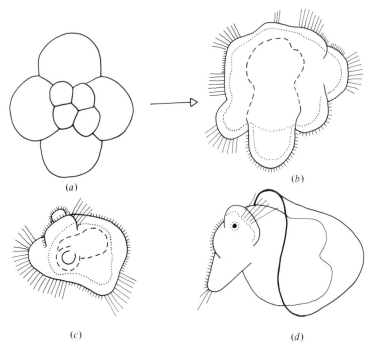

Fig. 5.6. Defect experiments on *Ilyanassa*. (*a*) An 8-cell stage embryo which had its polar lobe removed at the trefoil stage. It is quite symmetrical and there is no way of identifying a D macromere. (*b*) A larva arising from a lobeless embryo. It has velar cilia all over but lacks most of the normal structures and possesses little regional organisation. (*c*) A larva which has developed from an isolated AB blastomere. It is similar to the lobeless larva. (*d*) A larva formed from an isolated CD blastomere. This has some defects but is much more normal than the other types. (After Clement, 1952, 1956.)

alimentary canal other than the intestine are present although not organised as normal (Fig. 5.6). It is worth noting that the defects apparent in lobeless larvae are not simply of structures normally arising from the D lineage but of other parts as well (judged by the fate map assignments given in Fig. 5.2). This raises the possibility first suggested by Wilson (1929) that the D macromere which contains the polar lobe material is an organiser and is necessary for the normal development of the embryo because of some inductive signal which it emits.

To test this possibility, Clement (1962) removed the D macromere at different stages by puncturing it with a fine glass needle. The results are shown in a simplified form in Table 5.1. The main conclusion is that removal of the D macromere at the 4-cell stage produces a larva similar to the lobeless larva, lacking structures normally formed by the D lineage and also certain other structures such as the eyes and statocysts. But

Table 5.1. *Effects of removal of the D macromere of* Ilyanassa *at different stages*

Embryo	Defects expected from fate map	Defects found
ABC	Intestine, heart, shell, part of foot and gut	Intestine, heart, shell, foot, statocysts, eyes
ABC + 1d	As above	As above
ABC + 1d + 2d	Intestine, heart, part of foot and gut	As above
ABC + 1d + 2d + 3d	Intestine, heart, part of gut	Intestine, heart
ABC + 1d + 2d + 3d + 4d	Part of gut	None

removal of D after formation of the third micromere quartet deletes only those structures which normally arise from the mesentoblast itself. The implication is that the D macromere is required up to the 24-cell stage to control the patterning of the entire embryo, but that it can be dispensed with after this stage. Similar results have also been obtained with *Dentalium* (Cather & Verdonk, 1979).

A study of protein synthesis during the early cleavage stages has shown that isolated lobes and lobeless embryos make the same proteins as control embryos (Collier & McCarthy, 1981). After 24 hours, when the normal embryos have reached the 29-cell mesentoblast stage, the lobeless embryos still have nearly the same protein synthesis pattern as controls except for the persistence of two species which are normally turned off. This study is probably more reliable than previous ones since it uses two-dimensional rather than one-dimensional gels to resolve the proteins and it suggests that there is no differential localisation of maternal mRNAs in the polar lobe.

Although many polar lobes do not contain any special structures visible in the electron microscope, that of *Bithynia* contains a 'vegetal body' rich in RNA. This can be displaced by low-speed centrifugation and in contrast to *Ilyanassa* and *Dentalium* a proportion of embryos are unaffected by lobe removal if they have previously been centrifuged (Verdonk & Cather, 1973; Cather & Verdonk, 1974), suggesting that the vegetal body is associated with the polar lobe determinant.

Mollusc embryos are favourable to the experimentalist in several respects but particularly because single early blastomeres can be raised in isolation. According to our definitions in Chapter 2, a fragment which can self-differentiate certain structures *in vitro* is judged to be specified with respect to decisions in the developmental hierarchy which are necessary and sufficient conditions for the formation of those structures. Early isolation experiments were carried out by Crampton (1896) on *Ilyanassa* and Wilson (1904*a*, *b*) on *Dentalium* and *Patella*. These authors represented their results as showing a strict mosaicism, in other words that the states of specification corresponded exactly to the fate map. However, the results are open to certain criticisms. Firstly these authors did not know the fate map, indeed the situation is still unsatisfactory today, and without a good fate map it is difficult to decide whether or not regulation has occurred. Secondly, their isolates often became infected and did not survive very long. This meant that they were looking at early events such as the orientation of cleavages and the positions of cilia, but these are both features which will probably depend entirely on the cytoarchitecture of the egg and which will not be altered by any respecification of nuclei within the fragment.

More recent studies by Clement (1956) show that isolated blastomeres do not behave in a mosaic fashion. CD and D blastomeres can form larvae which are somewhat abnormal in form but contain most of the structures which normal larvae contain. On the other hand AB, A, B and C blastomeres develop into larvae resembling lobeless larvae which lack shell, foot, statocysts, heart and eyes (Fig. 5.6). It should particularly be noted that in normal development the eyes are probably formed from the 1a and 1c micromeres, although in isolated blastomere experiments D can develop eyes while A and C cannot. Likewise Cather (1967) showed that isolated A, B, C or D blastomeres could make shell while in normal development only 2d does so.

The implication is that at the 4-cell stage the A, B and C blastomeres are equivalent and that they require to undergo some interaction with the D lineage to acquire their definitive states of determination.

Determination of the D lineage in Patella

The results described up to now indicate that the organising properties of the D lineage arise from some cytoplasmic localisation which is established before the first cleavage and which is in some way associated with the polar lobe. However, some molluscs have no polar lobe; the cleavage is equal and it is not possible to say which of the macromeres is D until the mesentoblast has been formed. One such species which has been investigated recently is the common limpet, *Patella vulgata*.

According to the morphological description of van den Biggelaar (1977), two of the macromeres remain joined by a cross-furrow after the second cleavage and it is one of these which always becomes D. The first divisions of micromeres and macromeres are synchronous, so that a 32-cell stage is formed before any of the divisions producing the fourth quartet. The embryos remain at this stage for some time while the macromeres protrude up within the embryo to make contact with the apical micromeres of the molluscan cross. There then appears to be a struggle between the two cross-furrow-forming macromeres to see which can make the most intercellular contacts, and the one which does so becomes 3D. Deletion of one of the cross-furrow macromeres at the beginning of the 32-cell stage does not suppress formation of the mesentoblast although if one of the cells were preprogrammed to become D then the mesentoblast would be expected to be lost in 50% of cases. So in these embryos it seems that another macromere assumes a central position, makes the apical contacts and forms the mesentoblast. On the other hand the mesentoblast is not formed if all four macromeres are left in place but prevented from contacting the apical micromeres. This is done either by partial dissociation of the embryos in citrated sea water, or by deletion of the cells of the first quartet (van den Biggelaar & Guerrier, 1979).

Do the results on *Patella* indicate that this species has a wholly different mechanism of early development from the polar-lobe-forming types such as *Ilyanassa* and *Dentalium*? One is reluctant to accept this conclusion for reasons already discussed in Chapter 1 concerning homology and universality. It is possible, for example, that the localisation associated with the polar lobe is not a determinant which restricts the potency of the lineage but simply a bias in the cytoarchitecture which guarantees that the D macromere always makes the appropriate contacts and always forms the mesentoblast. On the other hand it is also possible that there is such a determinant and that it is made at the 1-cell stage in some species and at the 32-cell stage in others. It is the resolution of this type of question which is so important in informing a biochemical approach to early development, and only once such questions are answered shall we know where to look for the significant regulatory molecules.

Annelids

The early development of polychaete and oligochaete worms up to the trochophore stage is remarkably similar to that of molluscs and has caused zoologists to regard the two phyla as closely related despite the radical difference in morphology of the adults. A number of cell lineage studies were carried out around the turn of the century and as in molluscs showed the importance of the D lineage: the somatoblast, 2d, forms much of the trunk ectoderm and the mesentoblast, 4d, forms the mesodermal bands.

Some types of egg develop visible cytoplasmic specialisations near the animal and vegetal poles after fertilisation and some species possess a polar lobe. It is thus of interest to know whether these localisations are significant for cellular commitment. Low-speed centrifugation can be used to stratify the egg contents in the direction of centrifugal force but

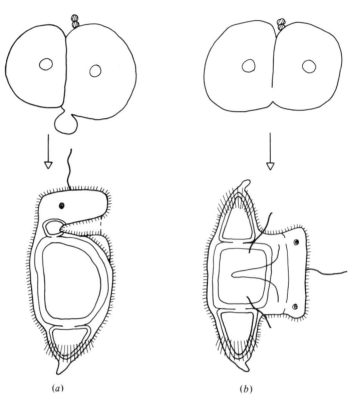

Fig. 5.7. Production of a mirror-duplicated larva of the polychaete *Chaetopterus* by compression at the first cleavage, distributing the lobe material equally between the first two blastomeres. (*a*) Normal trochophore, (*b*) Janus larva. (After Titlebaum, 1928.)

does not usually affect the course of development (Lillie, 1906; Morgan, 1927). However, if the cleavage pattern is altered by compression so that some polar lobe material enters both AB and CD blastomeres then mirror-symmetrical double embryos can be produced (Titlebaum, 1928; Tyler, 1930; Guerrier, 1970*b*; and Fig. 5.7). These bear an obvious resemblance to the double embryos produced in insects and amphibians by disturbance of the early cytoarchitecture which have already been discussed in Chapters 3 and 4, and suggest that the signal from the D macromere may, like those from the organiser and the activation centre, be some kind of gradient.

High-speed centrifugation can cause the egg to break into an anucleate yolky fragment and a nucleated yolk-depleted fragment, and the latter can develop into normally proportioned trochophores of reduced size (Wilson, 1929). It is also possible to obtain more or less normally proportioned larvae of reduced size by isolating the CD blastomere (Wilson, 1904c) or the D blastomere alone (Penners, 1926). In contrast the AB cell or A + B + C give rise to severely defective larvae.

Annelid embryos have not been the subject of much research using more modern techniques but this brief survey of classical results should indicate that in general the situation resembles that in molluscs. It is possible to reduce the scale of the pattern by various techniques but in order to produce a reasonably normal pattern the embryos need the polar lobe cytoplasm and the resulting D lineage.

Ascidians

Ascidians are sessile marine animals which often live in colonies and may be mistaken for plants by the uninitiated. They are classified along with the vertebrates in the phylum Chordata because they have free-living larvae called tadpoles whose morphology is obviously similar to that of vertebrates, with a dorsal neural tube, a notochord and segmented tail muscles.

Their embryonic development is exceptionally rapid. The eggs of the species considered here, *Styela* (*Cynthia*) and *Ciona*, are both 150 μm in diameter and develop to hatching tadpoles in only 18 hours. Some other species are even faster than this. With such a very rapid developmental tempo it is not surprising that developmental decisions are made very early with respect to morphological stage.

Normal development and fate map

The eggs are laid before maturation. The sperm enters near the vegetal pole and the polar bodies are extruded at the animal pole. After fertilisation there is some considerable rearrangement of the egg cytoplasm and in *Styela* a bright yellow crescent forms in the vegetal hemisphere, defining the prospective posterior end. Diagrams of the developmental stages are shown in Fig. 5.8 and the early cell lineage in Fig. 5.9.

The first cleavage is medial and bisects the yellow crescent. Because of the bilateral symmetry of the organism both of the first two blastomeres are called AB2. The second division is frontal, separating two anterior cells called A3 from two somewhat larger posterior cells called B3 which include the yellow crescent. The third cleavage is equatorial and separates four animal cells (a4.2 and b4.2) from four larger vegetal cells (A4.1 and B4.1). The synchrony of cleavages ceases by the 16-cell stage although the timing and orientation of each particular division are the same in all individual embryos.

The animal and vegetal cell layers continue to divide and a blastocoel forms between them. Gastrulation commences at the 64-cell stage as an

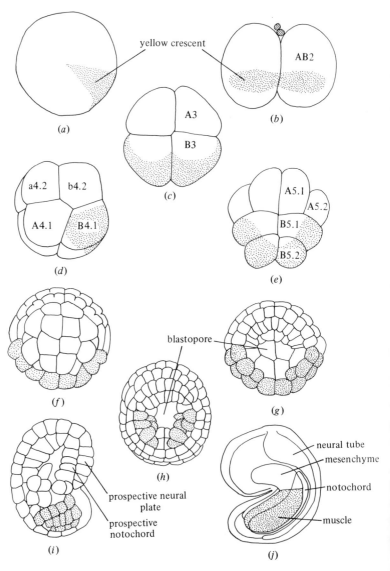

Fig. 5.8. Normal development of *Styela*. The pigment of the yellow crescent is shown stippled. (*a*) Fertilised egg, lateral view. (*b*) 2-cell stage, posterior view. (*c*) 4-cell stage, vegetal view. (*d*) 8-cell stage, lateral view. (*e*)–(*h*) Vegetal views, gastrulation commences in (*g*). (*i*) Parasagittal section through late gastrula. (*j*) Early tadpole, side view. (After Conklin, 1905a.)

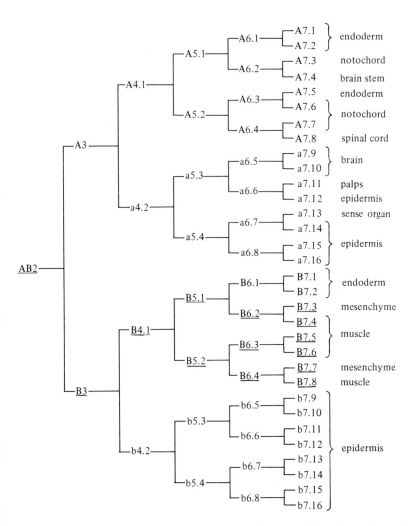

Fig. 5.9. Cell lineage of *Styela* up to the 64-cell stage. Since the embryo is bilaterally symmetrical, only one side is shown. The blastomeres which inherit the pigment of the yellow crescent are underlined. Fate map assignments according to Conklin (1905a) and Ortolani (1955).

invagination of the vegetal layer. At this stage most of the invaginating region is prospective endoderm but includes also the prospective notochord anteriorly and the prospective muscle and mesenchyme posteriorly. The animal layer, consisting of prospective neural plate and epidermis, spreads down and constricts the blastopore into a T shape with the bar anterior. As this T closes it is gradually moved posteriorly and the notochord and neural plate move round to what was the vegetal side and now becomes the dorsal side. Eventually the mesenchyme and muscle end up lateral to the notochord and the remnant of the blastopore has moved right round to the posterior. The neural plate forms a tube and sinks below the epidermis and the final result is a young tadpole with the basic chordate body plan.

The fate map is known with some precision as a result of the painstaking cell lineage study of Conklin (1905a), who followed every cell up to the 218-cell stage. This was possible because of the visibly different cytoplasmic zones set up after fertilisation which are passively parcelled out to the blastomeres. The yellow crescent is the most obvious of these but other regions can be distinguished as well by their content of yolk and pigment. Conklin's results have been slightly modified by Ortolani (1955) who constructed a fate map with the aid of chalk granules. The fate map assignments are shown in the diagram of cell lineage in Fig. 5.9. It is important to emphasise two things. Firstly the assignments are not clonal until quite a late stage. The boundaries of the prospective regions at early stages do not necessarily correspond to the cellular boundaries and so fate map assignments simply indicate in which structure most of the cytoplasm of a given blastomere will end up. Secondly, the fate map tells us nothing about cell commitment. This is not obvious from reading Conklin who evidently believed that the cytoplasmic regions were associated with determinants which caused the cells containing them to develop into the appropriate structures. But readers of this book will now be well aware that evidence about commitment must be obtained from other types of experiment.

Regional organisation

The evidence that we have suggests that there is unlikely to be much in the way of cytoplasmic determinants in the unfertilised egg. Centrifugation of unfertilised eggs has been shown to redistribute the plasms but not to inhibit normal development (La Spina, 1958). Dalcq (1932) found that half-size fragments resulting from meridional bisection of unfertilised eggs would develop normally after fertilisation while fragments resulting from equatorial bisection could give rise to partial embryos. Reverberi & Ortolani (1962), on the other hand, considered that reasonable twins could be produced from halves of the unfertilised egg arising from any orientation of cut. The difference between these results may perhaps be explained by the fact that the species used were different. These results indicate that in ascidians, as in other embryos, it is possible to produce normally proportioned larvae of reduced size after removal of parts of the egg. It is also possible to make double-size larvae by fusion of embryos at the 2-cell stage, providing that the fusion is carried out in such a way that the egg axes of the two embryos are aligned (von Ubisch, 1938). It thus seems that the sizes of the parts can adapt to the overall size of the organism over a 4-fold volume range. In some species there may be some cytoplasmic localisation before fertilisation while in others, judging from the visible cytoplasmic rearrangement, it is set up at fertilisation.

Deletion experiments on early cleavage stages were carried out by Chabry (1887) and Conklin (1905b). They concluded that if one of the first two blastomeres was killed, a lateral half-tadpole was formed from the remaining one, and if one or two cells of the 4-cell stage were killed then the larva lacked cells normally formed by those blastomeres. Given the very rapid tempo of ascidian development these mosaic results might be assumed to arise simply because there was not enough time for the patterning mechanism to adjust to the change of scale, particularly since the killed blastomeres remain attached.

Blastomere isolation experiments were carried out by Reverberi & Minganti (1946). They isolated the four pairs of equivalent cells from the 8-cell stage and found that the vegetal pairs formed tissues appropriate to the fate map but the animal pairs formed only ectodermal vesicles (Fig. 5.10). Brain and sense organs, which normally arise from the a4.2 pair, will only develop when these blastomeres are cultured in combination with the anterior vegetal pair A4.1. These authors regard the results as evidence for a neural induction which occurs in later embryonic development, as in amphibian embryos. It serves to illustrate the fact that the different states of specification which are inferred from isolation experiments correspond to the earliest decisions in the hierarchy and not to commitments to form particular terminally differentiated cell types.

Nevertheless it is possible to make some inferences about early decisions even if terminal events have to be used as end-points. An example of this is the detailed investigation of the formation of acetyl-cholinesterase in the embryos of *Ciona* and *Styela* carried out by Whittaker and coworkers (Whittaker, 1973, 1980; Whittaker, Ortolani & Farinella-Ferruzza, 1977). This enzyme normally appears specifically in the tail muscle at 8 hours of development which is the neurula stage. Its appearance is sensitive to actinomycin D up to 5 hours (gastrula) and to puromycin, which is a protein synthesis inhibitor, up to 7 hours. So we can deduce that normally transcription of the structural gene, or some other necessary gene, commences in the muscle lineage around 5 hours and that protein synthesis commences by 7 hours. Whittaker found that the

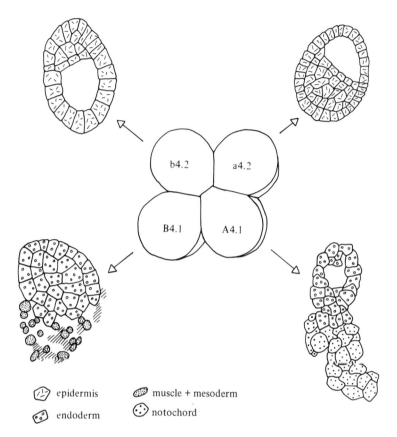

Fig. 5.10. Blastomere isolation experiments in an ascidian. Pairs of equivalent blastomeres from the 8-cell stage were cultured in isolation. They tend to produce the tissues which they form in normal development, except for the a4.2 pair which forms epidermis instead of neural tissue. (After Reverberi & Minganti, 1946.)

enzyme appeared on schedule in embryos whose cleavage and morphogenetic movements had been arrested from an early stage by cytochalasin B. It appeared only in cells judged from the fate map to have been prospective muscle cells at the time of the cleavage block, and moreover the times of actinomycin D and puromycin sensitivity were the same as normal (Fig. 5.11). The synthesis of a terminal differentiation product in specific cleavage-blocked blastomeres has also been demonstrated in embryos of the nematode *Caenorhabditis* (Laufer, Bazzicalupo & Wood, 1980).

This shows that the 'clock' controlling gene activation does not depend on the cell division cycle. But it does not in itself show that a determinant for muscle was present in a certain region of the embryo from the 2-cell

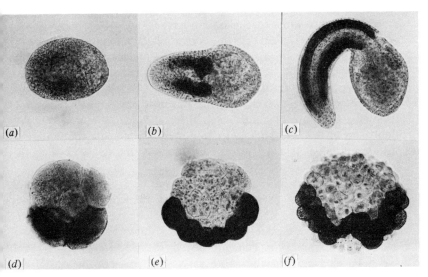

Fig. 5.11. Development of acetylcholinesterase in the ascidian *Halocynthia*. (*a*)–(*c*) Normal development: the enzyme is demonstrated by a histochemical reaction which shows that it is localised in the tail muscle. (*a*) Neurula, (*b*) tailbud, (*c*) tadpole stage. (*d*)–(*f*) Embryos cleavage-arrested by cytochalasin B. (*d*) Arrested at 8-cell stage, (*e*) at 32-cell stage, (*f*) at late gastrula. The enzyme appears on schedule in those cells destined to contribute to the tail muscle. (Photographs kindly provided by Dr N. Satoh.)

stage, since the nucleus and cytoplasm of the blastomeres which produce the enzyme may go through many distinct states of commitment between the cleavage block and the onset of transcription.

In further experiments, Whittaker and coworkers have isolated the B4.1 blastomere pair from the 8-cell stage and cultured them until control embryos reached the tailbud stage. Acetylcholinesterase develops in the isolates and not in the residual embryos which lacked B4.1. This indicates that interactions with the remainder of the embryo after the 8-cell stage are not necessary for formation of the enzyme.

Finally Whittaker has produced an abnormal distribution of cytoplasm at the 8-cell stage of *Styela* by compressing embryos at the 4-cell stage. This prevents the usual equatorial cleavage and means that four cells inherit the yellow crescent cytoplasm instead of the usual two. These pressed embryos develop abnormally after release of pressure. However, if they are cleavage-blocked by cytochalasin B a significant minority produce acetylcholinesterase in all four of the blastomeres containing yellow crescent material while cleavage-blocked controls form the enzyme only in the B4.1 pair. This seems a clear case of nuclear reprogramming by a cytoplasmic factor, similar to the pole plasm graft in *Drosphila* (Chapter 4). It is unlikely that the yellow pigment itself is a determinant since some species do not have it, and its redistribution by centrifugation of fertilised eggs does not affect development. Also many cells of the crescent lineage normally become mesenchyme rather than muscle. However, it seems fairly clear that at some time between fertilisation and the fourth cleavage a factor becomes localised in the yellow crescent region which is capable of causing associated nuclei to set off down a sequence of states which in the absence of interactions with other parts of the embryo lead to the formation of acetylcholinesterase some hours later. It would certainly be interesting to know what this factor is and how it works.

The role of the cytoplasmic determinant has been further investigated by Satoh and Ikegami (Satoh, 1979; Satoh & Ikegami, 1981a, b). They used mainly the species *Halocynthia roretzi* which develops a little slower than *Styela* but whose acetylcholinesterase shows the same stages of sensitivity to actinomycin D and puromycin. As in *Styela*, embryos which have been cleavage-blocked at an early stage by cytochalasin B will produce the enzyme at the usual time in the blastomeres of the muscle lineage. Cytochalasin B binds to microfilaments and accordingly inhibits cytokinesis and cell movements but allows DNA replication and nuclear division to continue. But when the DNA synthesis inhibitor aphidicolin is added to early stages cleavage is inhibited after one more cell division and the blocked embryos do not then express the enzyme. If the aphidicolin is added at the 76-cell stage or shortly afterwards, which is still before the actinomycin-sensitive period, the enzyme later appears in some but not all of the cells of the muscle lineage. Satoh & Ikegami (1981a) showed that the cells in which the enzyme appeared were those which had undergone seven rounds of DNA replication by the time the aphidicolin was added. It is possible to know this because the cell division is markedly asynchronous by this stage. In the early gastrula some prospective muscle cells are in the seventh generation (B7.5, B7.6), some in the eighth (B8.15, B8.16), and some in the ninth (B9.13, 9.14, 9.15 and 9.16). They suggest that DNA replication provides a developmental clock and after seven rounds of replication (eighth cell generation) the nuclei become competent to interact with the cytoplasmic determinant and become committed to produce the enzyme. It may be significant that it is at the cell division before this stage that the prospective muscle lineage becomes clonal, with the division of B6.2 and B6.4 each into one muscle and one mesenchyme precursor cell.

These experiments are interesting because they distinguish between different kinds of developmental commitment. The cells are specified, according to the test of isolation, by the 8-cell stage, but the commitment to form muscle is not clonally heritable before the sixth cleavage (cellular determination) and the nuclei are not committed to autonomous expression of acetylcholinesterase until one cleavage later than this (nuclear determination).

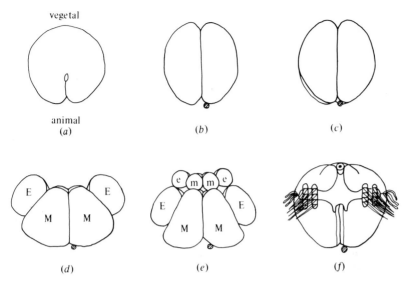

Fig. 5.12. Normal development of a generalised ctenophore: (*a*) first cleavage, (*b*) 2-cell stage, (*c*) 4-cell stage, (*d*) 8-cell stage, (*e*) 16-cell stage, (*f*) cydippid larva. In all figures the vegetal pole is up and the animal pole down. (After Freeman, 1977.)

Ctenophores

Ctenophores are marine planktonic organisms which are notable for their ability to produce light. They are biradially symmetrical, having two planes of symmetry, the sagittal and the tentacular, which partition the organism into four equivalent quadrants (Fig. 5.12). At right angles to these planes runs the principal body axis, the oral–aboral axis. At the aboral pole is an apical organ from which run eight rows of ciliated comb plates. Below these lie radial canals containing the gonads and the light-producing cells or photocytes: small round cells containing the photoprotein aequorin which emits light when treated with calcium ions.

Although quite unlike ascidians in morphology, ctenophores also develop very rapidly, motile cydippid larvae forming in 10 hours in some species, and they also provide favourable material for the study of cytoplasmic localisation at early stages.

Normal development and fate map

The eggs of different species vary somewhat in diameter (*Beroe* 1 mm, *Mnemiopsis* 150 μm) but the sequence of early cleavages is the same (Fig. 5.12). The polar bodies are given off at the animal pole after fertilisation although since ctenophore development appears to be 'upside down' compared with other animals it is possible that their animal pole is really homologous to the vegetal pole elsewhere. The first cleavage occurs in the sagittal plane and the second in the tentacular plane. The third cleavage is oblique producing four internal M blastomeres and four smaller external E blastomeres. Each of these then undergoes an unequal cleavage to form a micromere at the vegetal side. The micromeres derived from M are called m and those derived from E are called e. In *Beroe* the outer layer of the unfertilised egg consists of a green fluorescent 'ctenoplasm' rich in mitochondria. On fertilisation this concentrates at the vegetal pole and at the fourth cleavage it is segregated into the eight micromeres.

The micromeres themselves divide and are joined by further octets from the macromeres. Gastrulation consists of an epibolic spreading of the micromeres and a turning in of the macromeres to form the oesophagus. The apical organ appears at the vegetal pole and four ectodermal thickenings radiate from it, each becoming a paired row of comb plates. The radial canals develop somewhat later from endodermal pockets. In *Mnemiopsis* gastrulation occurs at about 3 hours, the first comb plate cilia appear around 9 hours, and the larvae will hatch at 30–36 hours at 19°C.

Reverberi & Ortolani (1963) studied the fate of the blastomeres of the early stages by applying coloured chalk granules. They showed that marks applied at the animal pole invaginated during gastrulation and ended up in the gastric cavity while marks on the vegetal pole ended up near the apical organ at the aboral pole. Marks on the e micromeres labelled the ciliated combs, on the m micromeres the apical organ, the E macromeres the radial canals and the M macromeres the gastric sac (Fig. 5.13). Deletion experiments by Freeman & Reynolds (1973) suggest that the photocytes are produced from micromeres m3 which are later derivatives of M.

Regional organisation

Both unfertilised and fertilised eggs will give rise to normally proportioned larvae of reduced size after division into two parts (Yatsu, 1911, 1912). Centrifugation before the first cleavage can alter the normal relationship of the cleavage plane to the polar bodies but does not affect subsequent development (Freeman, 1977).

However, isolated blastomeres always produce partial larvae. Blastomeres from the 2-cell stage give larvae with four rows of comb plates (Driesch & Morgan, 1895). Isolated blastomeres from the 8- and 16-cell stages were studied by Farfaglio (1963) who showed that each E blastomere would form a defective larva with two comb rows. Even isolated e micromeres would produce some irregular comb plates while whole embryos with all four e cells deleted lacked comb plates altogether.

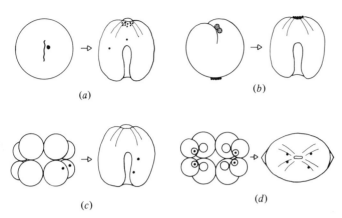

Fig. 5.13. Fate mapping of the early ctenophore embryo. (*a*) Marks on the animal pole end up in the gastric cavity. (*b*) Marks on the vegetal pole end up at the aboral pole. (*c*) Marks on the E macromeres end up in longitudinal canals under the comb rows and marks on the M macromeres in the gastric sac. (*d*) Marks on the e micromeres end up on the ciliated combs. (After Reverberi & Ortolani, 1963.)

The development of luminescence was studied by Freeman & Reynolds (1973). They measured low levels of light production elicited from embryos by treatment with potassium chloride, which increases membrane permeability and allows ingress of calcium ions. The first photocyte activity in normal development of *Mnemiopsis* was detected around the time of comb plate formation. Isolated M macromeres could produce photocytes at around the same time, while embryos in which all four M cells had been deleted produced no photocytes.

So by the 16-cell stage the blastomeres have been differentially specified to the extent that the comb-plate-forming potential lies in the e micromeres alone and the photocyte-forming potential in the M macromeres alone. The segregation of comb plate potential and its relation to cleavage has been studied in some detail by Freeman (1976*a*, *b*, 1977).

Removal of aboral cytoplasm from fertilised eggs, cleaving and 2-cell stages produced only a slight reduction in comb plate formation. This suggests that there is little or no localisation of comb-plate-forming potential along the oral–aboral axis by the 2-cell stage, although only the e micromeres, which are on the aboral side, have it by the 16-cell stage.

When they are formed the e micromeres are external as well as aboral in position. In a further series of experiments the internal or external cytoplasm was removed from a 2- or 4-cell stage blastomere *in situ* and then the E and M cells arising from the defective blastomere were isolated and allowed to develop as an assay for comb plate or photocyte formation (Fig. 5.14). When the operation was done at the 2-cell stage isolated blastomeres behaved as normal: the E cells formed comb plates and the

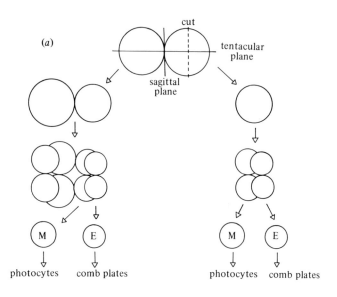

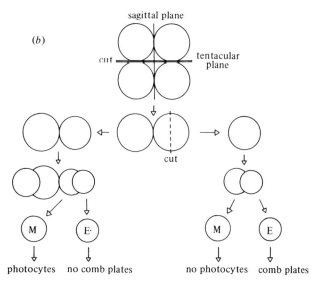

Fig. 5.14. Experiment to show that cytoplasmic localisation in ctenophore embryos occurs shortly before the stem cell division. In (*a*) a blastomere from the 2-cell stage is cut in half. Only the nucleated fragments continue development and these may be internal or external halves in different individual embryos. They are picked out and the resulting E and M blastomeres are separated and allowed to develop in isolation. Despite containing different regions of the egg cytoplasm from normal they self-differentiate the usual structures. In (*b*) the experiment is the same except that the cut is made at the 4-cell stage. Here the 'outer' M and the 'inner' E blastomeres do not form the usual structures, implying that cytoplasmic localisation must have occurred between the 2- and 4-cell stages.

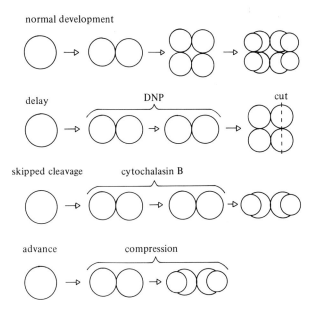

normal development

delay   DNP   cut

skipped cleavage   cytochalasin B

advance   compression

Fig. 5.15. Experiments to show that localisation of the comb plate determinant in *Mnemiopsis* is not retarded by a reversible second cleavage block, nor advanced by compression.

M cells formed photocytes. But when the operation was done at the 4-cell stage the isolates were defective. E cells arising from blastomeres whose external cytoplasm had been removed failed to form comb plates and M cells from blastomeres whose medial cytoplasm had been removed failed to form photocytes. This suggests that if there is some segregation of determinants along the tentacular axis then it probably occurs at the 4-cell stage, i.e. during the cell cycle before the unequal cleavage.

This conclusion is reinforced by experiments in which the timing and geometry of cleavage are altered (Freeman, 1976*a, b*). It is possible to cause the embryos to skip one or more cleavages by treating them with dinitrophenol (DNP) or cytochalasin B. The clearest results are those obtained after suppression of the second cleavage. The treated embryos may simply continue their program, delayed relative to controls, or they may proceed to the normal configuration with half the number of cells (Figs. 5.15 and 5.16). The first type is more common with DNP and the second with cytochalasin. A delayed 4-cell embryo can be shown to have a localisation of potential within its blastomeres, since although the isolated blastomeres form both comb plates and photocytes the potency to form either can be reduced by removing the appropriate regions of cytoplasm. Likewise a 4-cell embryo of 8-cell configuration has E and M blastomeres which behave differently on isolation as do blastomeres from

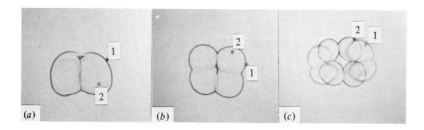

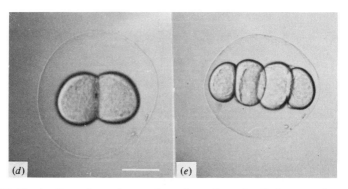

Fig. 5.16. Production of an abnormal configuration of cells in an embryo of *Mnemiopsis*. (*a*)–(*c*) 2-, 4- and 8-cell stages in normal development. The chalk granules labelled 1 and 2 indicate the cytoplasmic regions which normally enter the E and M cells. (*d*) and (*e*) 2- and 4-cell stages after suppression of the second cleavage. In (*e*) the embryo has adopted an '8-cell' configuration with only four cells. (Photographs kindly provided by Dr G. Freeman.)

8-cell controls. So localisation seems to occur at the time of the normal 4-cell stage irrespective of whether two or four cells are actually present.

By contrast, when a normal 2-cell embryo is compressed so that the second cleavage is sagittal the four blastomeres are all equivalent and form both comb plates and photocytes on isolation, despite the fact that they have the same configuration as an 8-cell stage. So localisation cannot be advanced by artificially altering the planes of cleavage, nor can it be retarded by reversible cleavage blocks. Evidently localisation is controlled by its own clock and under normal circumstances this guarantees that the unequal third cleavage will produce cells of different specification.

It should perhaps be mentioned that the interpretation of Freeman's results presented here is not necessarily the same as that of Freeman himself. For this the original papers must be consulted.

General remarks on 'mosaic' embryos

In the early years of the present century there was believed to be a fundamental difference between regulative embryos, in which twinning was possible, and mosaic embryos, in which parts developed according to the fate map. But it was soon recognised that there was always a mosaic stage even if it barely preceded terminal differentiation and so the conception arose that mosaic embryos were simply precocious in their developmental decision-making (see, for example, discussion in Huxley & de Beer, 1934). This view has been supported by experimental work which has shown that a good deal of regulative behaviour is in fact always found at the very earliest stages.

We have seen that in all the forms considered the overall size of the embryo can be altered; sometimes because a single blastomere can develop to form a complete miniature larva, but more usually because some regions of cytoplasm can be removed from the egg before fertilisation. Moreover stratification of the egg contents by centrifugation rarely perturbs the course of normal development.

On the other hand determinative regions of cytoplasm clearly exist. The most direct evidence is provided by the cases in which the distribution of a particular cytoplasmic region is artificially altered and this brings about an abnormal pathway of development for the cells into which it has been introduced. This is shown by the compression experiment of Teitlebaum (1928) in which polar lobe cytoplasm enters both of the blastomeres of an annelid 2-cell stage and a Janus-larva results. It is also shown by Whittaker's (1980) compression experiment on *Styela* in which acetylcholinesterase is formed by blastomeres containing yellow crescent cytoplasm. It is also shown, of course, by the pole plasm graft in *Drosophila* (Illmensee & Mahowald, 1974). So while mosaic behaviour of parts of embryos is not in itself evidence for a mosaic of cytoplasmic determinants in the egg, it is clear that some cytoplasmic determinants do exist, sometimes in the unfertilised egg, sometimes after fertilisation, and sometimes within particular blastomeres at later stages.

On the whole two conclusions about these determinants seem probable, although not certain, given the present state of the evidence. Firstly, it does not seem as though they code for terminally differentiated cell types even though these cell types may serve as the markers in experimental work. The reason for believing this is that in normal development the cells containing particular plasms always enter more than one tissue. The pole cells of *Drosophila* become part of the mid-gut as well as becoming germ cells; the yellow crescent of *Styela* enters mesenchyme as well as muscle; and the polar lobe cytoplasm of annelids and molluscs enters a variety of cell types formed by the D lineage. Since further decisions are clearly necessary to bring about the differentiated cell types the determinants presumably code for early choices in the hierarchy.

Secondly, it seems likely that in each case the regionalisation occurs shortly before the asymmetrical cell division which generates the differently specified cells. In the ctenophores the 'comb plate determinant' is localised only by the 4-cell stage and the differentially specified E and M lineages are set up at the next cleavage. In *Styela* the yellow crescent arises after fertilisation and the second cleavage is asymmetrical. In those molluscs with an asymmetrical first cleavage the polar cortical cytoplasm is presumably laid down during oogenesis, while in those with symmetrical first cleavage the D cell arises later as a result of some interaction with the apical micromeres.

So what we are dealing with is probably not a 'mosaic' of determinants coding for many differentiated cell types all over the body but only one determinant which in each case becomes localised shortly before the critical cell division. When the phenomenon is expressed in this way it is apparent that it does not necessarily reflect a totally different type of biochemical mechanism from the cases in which whole cell sheets appear to become regionalised in response to inductive signals. The scale of size is similar in both cases (100 μm – 1 mm) and the only difference would appear to be that in one case cell membranes form before regionalisation while in the other they do not. In either case differential gene expression probably occurs somewhat later than cellular specification.

Sea urchins

Sea urchins share with amphibians the status of 'senior citizens' in embryological research, for it was with them that it was first shown that normally proportioned larvae of reduced size can be produced from embryo fragments (Driesch, 1891). Recently sea urchin embryos have been used a great deal for biochemical studies (Davidson, Hough-Evans & Britten, 1982) although unfortunately few of these data are relevant to the problems of the present book because not many experiments have been carried out on the different regions of the embryo. For obvious technical reasons workers have preferred to study DNA, RNA and protein synthesis in whole embryos at different stages rather than in different regions of the embryo at the same stage.

Normal development and fate map

A variety of species have been used for experimental work and some of their individual peculiarities are mentioned by Hörstadius (1973). The eggs are 75–150 μm in diameter, they mature in the ovary before being shed and are surrounded by jelly coats. After fertilisation a membrane is formed by combination of the vitelline membrane with material released from cortical granules. The first two cleavages are vertical and the third is equatorial (Fig. 5.17). In the fourth cleavage the four cells at the animal pole cleave equally to give eight mesomeres, while the four at the vegetal pole cleave unequally to form four macromeres and four micromeres. As cleavage continues, a blastocoelic cavity develops in the interior and long stereocilia form at the animal pole.

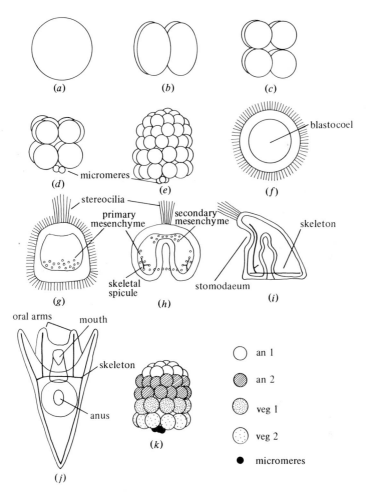

Fig. 5.17. Normal development of the sea urchin *Paracentrotus lividus*. (*a*)–(*e*)
cleavage and segregation of micromeres, (*f*) blastula, (*g*) early gastrula, (*h*) late
gastrula, (*i*) prism stage, (*j*) pluteus larva, (*k*) 64-cell stage shaded to show
nomenclature of the different cell layers. (After Hörstadius, 1935.)

The micromeres invaginate at the vegetal pole to form the primary
mesenchyme which later produces the skeleton. Gastrulation proper
consists of an invagination of the remaining vegetal tissue to form an
archenteron. The future anterior side flattens and forms a stomodaeal
ingrowth which fuses with the invagination to give the alimentary canal.
The cells at the tip of the invagination are known as secondary mesenchyme and form muscle and echinochrome pigment cells. Simultaneously
the embryo elongates at all extremes to form four projections – the two
oral arms, the acron and the apex – and these are reinforced by skeletal

rods laid down by the primary mesenchyme. By this stage the embryo has developed into the pluteus larva which is the usual end-point for embryological experiments. The adult sea urchin is produced somewhat later by a complex metamorphosis (described by Okazaki, 1975).

A fate map was constructed for *Paracentrotus lividus* by Hörstadius (1935, 1939) using vital stain marks applied to single blastomeres, and also using the subequatorial pigment ring which forms after maturation in the eggs of this species. He showed that the primary mesenchyme arises wholly from the micromeres and that the archenteron and secondary mesenchyme come from the ring of macromeres abutting them (called veg 2 in the 64-cell stage: Fig. 5.17*k*).

The cleavage stages of sea urchin embryos continue in actinomycin D concentrations sufficient to inhibit most of the RNA synthesis, although gastrulation does not take place. The rate of protein synthesis is unaffected by the inhibitor and this is therefore presumed to be directed by maternal mRNA (Gross & Coustineau, 1964). Maternal messages have been of great interest to biochemical embryologists in recent years (see Davidson, 1976) and are known to exist in masked form as ribonucleoprotein particles (Spirin, 1966; Gross *et al.*, 1973) most of which become active after fertilisation. The qualitative pattern of protein synthesis was studied using two-dimensional gel electrophoresis by Brandhorst (1976), who concluded that there were no significant changes at fertilisation and few before gastrulation.

Regional organisation

In common with other invertebrates, the eggs of sea urchins have been centrifuged, either at low speed to stratify the cytoplasm or at higher speed to produce fragments of different cytoplasmic composition. Following fertilisation stratified eggs develop quite normally (Lyon, 1906). Fragmentation experiments were conducted by Harvey (1933) who broke *Arbacia* eggs into 'red' and 'white' halves, the former containing yolk and pigment and the latter the nucleus, soluble cytoplasm and lipid. Both halves are capable of forming small plutei after fertilisation although of course the red larva is haploid and the white larva diploid.

Blastomere isolation experiments have shown that both of the first two blastomeres and all of the first four blastomeres can develop to form miniature plutei (Boveri, 1908). From starfish embryos it is possible to produce identical octuplets from the 8-cell stage (Dan-Sohkawa & Satoh, 1978; Fig. 5.18). These miniature embryos develop with the normal developmental tempo but at any given stage have a half, a quarter or an eighth the number of cells possessed by controls, depending on the sizes of the isolates. With sea urchins it is also possible to produce giant plutei by fusing together two 32-cell stages in such a way that their animal–vegetal axes are aligned (Hörstadius, 1957). So in both starfish and sea

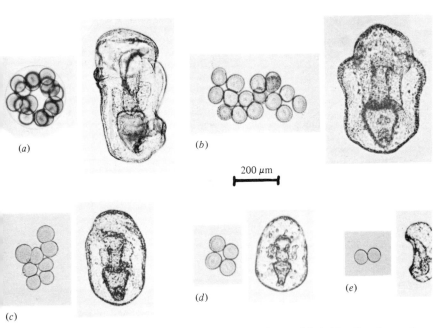

Fig. 5.18. Regulation of proportions in starfish embryos. (*a*) A 16-cell embryo of *Asterina pectinifera* and a normal bipinnaria larva. (*b*) The same except that the embryo was removed from the fertilisation membrane. (*c*) After removal of half the cells a half-size larva develops. (*d*) A ¼ embryo yields a quarter-size larva. (*e*) An ⅛ embryo yields an eighth-size larva. (Photographs kindly provided by Dr M. Dan-Sohkawa.)

urchins size regulation is possible over at least an 8-fold volume range, corresponding to a 2-fold linear range, and probably more.

In sea urchins after the 4-cell stage it is still possible to produce twins by meridional separation, but separation of fragments along the animal–vegetal axis usually produces defective plutei (Hörstadius, 1935; review 1939, 1973). Animal halves tend to produce ciliated balls, sometimes containing a stomodaeum, while vegetal halves tend to produce 'ovoids' lacking stereocilia, oral arms and stomodaeum (Fig. 5.19). However, the embryo is able to form a complete pattern following minor defects in the egg axis: it is possible to remove the an 1 ring of cells (see Fig. 5.17*k*) or the micromeres, or to combine an animal half with two or more micromeres (i.e. remove veg 1 and veg 2: Fig. 5.17*k*). This indicates that the regional pattern along the egg axis depends on an interaction which occurs at or after the 32-cell stage. Although there does not seem to be any specific level of tissue which is essential for establishment of the pattern it may be significant that it is possible to remove more tissue from the middle than from the extreme positions. It has often been suggested that the micromeres are an organising centre since they can induce a secondary

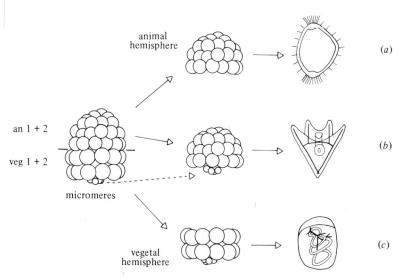

Fig. 5.19. Evidence for an inductive interaction along the axis of the sea urchin embryo. The isolated animal hemisphere forms a ciliated ball (*a*), the isolated vegetal hemisphere forms an 'ovoid' (*c*), but the animal hemisphere in combination with the micromeres forms a reasonably well-proportioned pluteus (*b*).

archenteron if implanted at unusual sites in the vegetal hemisphere (Hörstadius, 1935). If they are indeed the source of a gradient-type signal in normal development it is evident that this function can be taken over by the neighbouring veg 2 cells following their removal. The significance of this will be considered in Chapter 9.

Animalisation and vegetalisation

If whole embryos are treated with certain chemical substances they develop as though sufficient tissue had been removed from the vegetal or the animal end to produce a defect ('animalisation' and 'vegetalisation' respectively: Lallier, 1975). The best-known animalising agent is isocyanate and the best-known vegetalising agent is lithium ion. The biochemical basis of their activity is not known although it has recently been shown that there is no difference between the two-dimensional gel patterns of protein synthesis in normal and vegetalised embryos (Hutchins & Brandhorst, 1979). This suggests either that the transcription of proteins of low abundance is being affected, or that gene expression is not being affected at all and the activity of lithium is at the level of intermediary metabolism. Some protein fractions have been extracted from sea urchin embryos which have animalising and vegetalising activity (Runnstrom, 1975). It is of some interest that lithium ion also produces pattern abnormalities in amphibian embryos; if given during cleavage stages it will vegetalise and if given during gastrulation it will suppress the head and sometimes the notochord (see Chapter 3). It is possible that this is coincidental but it may indicate a similarity in biochemical mechanism of the first inductive interactions in sea urchin and amphibian embryos.

General references

Reverberi, G. (1971). *Experimental Embryology of Marine and Freshwater Invertebrates.* North-Holland, Amsterdam.
Hörstadius, S. (1973). *Experimental Embryology of Echinoderms.* Clarendon Press, Oxford.
Subtelny, S. & Konigsberg, I. R. (eds.) (1979). *Determinants of Spatial Organization.* Academic Press, New York & London.

6

The mouse and its amniote relatives

Compared with the subjects of the last three chapters mammalian experimental embryology is a relatively new subject, and for this reason it has a rather distinctive flavour. There are three components to this difference. First, grafting experiments have often been carried out with single cells rather than with pieces of tissue. Secondly, a determined effort has been made using biochemical and immunological methods to detect markers of the different regions at early developmental stages. Thirdly, there is the use of teratocarcinoma cells, which can be grown in large quantities *in vitro*, as 'model embryos'.

The main reason for the delay of serious experimental work on mammalian embryos until the 1950s was technical. Mammals are viviparous and the embryos cannot survive in isolation without the use of techniques which grew out of tissue culture. For the early, pre-implantation, stages the embryos are located first in the oviduct and then the uterus of the mother. During this time they can be collected and kept *in vitro* in reasonably simple media consisting of a buffered salt solution containing some protein and a respiratory substrate, and most experiments have involved manipulation of these early stages. Where later development is essential to the result the pre-implantation embryos are reimplanted into the uteri of 'foster mothers' who have been made 'pseudopregnant', and thus receptive to the embryos, by previous mating with sterile males.

The culture of post-implantation stages *in vitro* is an exacting business. With mouse embryos the best results to date have involved culture to early somite stages in media supplemented with foetal calf serum and human placental cord serum (Hsu, 1980). No doubt it will eventually be possible to raise mammalian embryos through the entire gestation period by artificial means, although one hopes that this will not be too soon in view of the possible application to human embryos. Up to the present, however, mammalian embryology is really 'pre-embryology', since it deals with the formation not of the embryo body plan but of various extracellular membranes which are segregated during the pre-implantation phase and which are necessary to the support and nutrition of the embryo proper.

The embryos of mammals, birds and reptiles are unusual in the animal kingdom in that they undergo a considerable amount of growth during development. Most other types of embryo develop in isolation from the mother and if they grow at all do so only to a relatively small extent at the expense of the yolk in the egg. By contrast, mammalian embryos after implantation obtain nourishment from the maternal blood stream, and birds and reptiles lay very large eggs containing a vast amount of yolk which is gradually used to nourish the foetus.

The extensive growth and the associated early formation of extra-embryonic structures have two important implications for the embryologist. The first concerns the question of proportion regulation. We have seen that in all embryos considered so far the size of structures depends on the size of the whole embryo. For example, in animals where twinning is possible following separation of the first two blastomeres, the twins are of normal proportions but half normal size. As we shall see this is not the case for mammals. In the pre-implantation stage there does not appear to be an exact regulation of the proportions of embryonic to extra-embryonic tissue. At later stages the extensive growth allows for adjustment of size and so twins become approximately normal-sized well before the end of the gestation period.

The second is the question of the relation between growth and the successive steps of the developmental hierarchy. If it is possible to disengage growth and decision making then it should be possible to cultivate cells in particular states of determination for indefinite periods. This offers a new dimension of opportunity for the embryologist for it provides access to quantities of material sufficient for biochemical work. Many investigators think that teratocarcinoma cells are indeed arrested in an early determined state and although it is at present difficult to be sure of their true affinities these should become clearer as biochemical characterisation becomes more complete.

The mouse embryo

Normal development

Ovulation in the mouse occurs a few hours after mating and fertilisation takes place at the upper end of the oviduct. The second polar body is expelled 2–3 hours later. The egg is surrounded by a vitelline membrane and a thick zona pellucida composed of mucopolysaccharide secreted by the ovarian follicle cells. The total diameter including the zona is around 100 μm, increasing somewhat after fertilisation.

The course of normal development up to the primitive streak stage is shown in Fig. 6.1. Development through the cleavage stages is very slow and in contrast to the embryos considered in the previous chapter there is

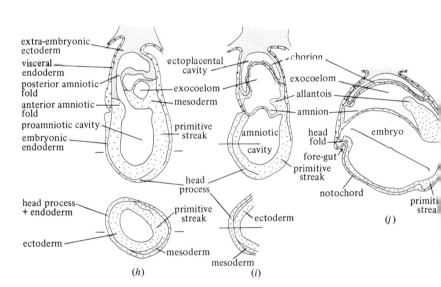

Fig. 6.1. Normal development of the mouse. (*a*) Zygote, (*b*) 2-cell stage, (*c*) 8-cell stage, (*d*) 8-cell stage after compaction, (*e*) blastocyst, (*f*) implantation stage, (*g*) egg cylinder, (*h*)–(*j*) development of embryonic membranes. The embryo does not grow before implantation but from (*f*) to (*j*) would increase in size by a factor of about six in linear dimensions. (Adapted from Snell & Stevens, 1966, except (*d*).)

evidence that expression of the paternal genome commences very early (Chapman *et al.*, 1976). The first cleavage occurs about 24 hours after fertilisation and the second and third cleavages, which are not entirely synchronous, follow at intervals of about 12 hours. In the early 8-cell stage the individual cells are still apparent but they cease to be visible when the whole embryo acquires a spherical shape in a process called compaction. Compaction is inhibited by removal of calcium or addition of cytochalasin B and is associated with the formation of tight junctions between the blastomeres (Ducibella & Anderson, 1975). Gap junctions are also formed at this stage and allow electrical coupling and diffusion of low-molecular-weight dyes throughout the embryo (Lo & Gilula, 1979).

The embryo is called a morula from compaction until about the 32-cell stage when a fluid-filled blastocoel begins to form in the interior. This is about 3 days after fertilisation and around the time that the embryo moves from oviduct to uterus. The activity expands the embryo into a blastocyst which consists of an outer cell layer called the trophectoderm and a clump of cells attached to one point of its interior, the inner cell mass (ICM). At the 60-cell stage, about one quarter of the cells are found in the ICM and three quarters in the trophectoderm. As we shall see, the ICM and trophectoderm are stably different cell types and biochemical differences can be demonstrated between them (Van Blerkom, Barton & Johnson, 1976).

From 3½ to 4½ days post-fertilisation both the ICM and the trophectoderm diversify. The ICM delaminates a layer of 'primitive endoderm' on its blastocoelic surface. Despite its name this layer is not now thought to contribute to the embryonic 'definitive' endoderm but rather to extra-embryonic membranes. The trophectoderm becomes divided into a polar component, which overlies the ICM, and a mural component which makes up the remainder. While the polar trophectoderm continues to proliferate, the mural trophectoderm becomes transformed into polyploid giant cells in which the DNA continues to be replicated but without mitosis.

At about this stage the embryo hatches from the zona and becomes implanted in a uterine crypt. The trophectoderm then becomes known as trophoblast and soon stimulates proliferation of the connective tissue of the uterine mucosa to form a decidual swelling. The uterus is not competent to receive the embryo except during a short period about 5 days from copulation – hence the use of pseudopregnant females as foster mothers in experimental work. From this stage onward the embryo is in a position to grow in size and weight; during the pre-implantation stage it has been losing dry weight because it must find from within much of the material for its own metabolic activity.

The next stage of development is known as the egg cylinder. Cells derived from the primitive endoderm move out to cover the whole inner surface of the mural trophectoderm and start to secrete an extracellular basement membrane known as Reichert's membrane which contains type IV collagen, proteoglycan and laminin (Hogan, Cooper & Kurkinen, 1980). These cells are called the parietal endoderm. The remainder of the primitive endoderm remains epithelial and forms a layer of visceral endoderm around the egg cylinder. The cells of this layer are characterised by the synthesis of α-foetoprotein and transferrin (Dziadek, 1979). The inside of the egg cylinder consists of the 'ectoderm' from which the entire embryo will develop, and extra-embryonic ectoderm derived from the polar trophectoderm. This extra-embryonic region projects above the outer surface of the trophoblast as the ectoplacental cone and as it proliferates it produces further layers of giant cells which move around and reinforce the trophoblast.

At about 6 days the anteroposterior axis of the future embryo becomes apparent with the formation of the primitive streak at the posterior end of the embryonic ectoderm. The streak is a locus of cell movements which are not well understood but result in the formation of the definitive endoderm and mesoderm in the usual trilaminar arrangement. The amniotic fold forms as an outpushing of the ectoderm and mesoderm at the junction of primitive streak and extra-embryonic ectoderm. The side of this fold nearer the embryo becomes the amnion and the side nearer the ectoplacental cone becomes the chorion. The fold pushes across the proamniotic space and divides it into three: an amniotic cavity above the embryo, an exocoelom separating amnion and chorion, and an ectoplacental cavity lined with extra-embryonic ectoderm. By 7 days a head process appears in the definitive endoderm between the anterior extremity of the primitive streak and the anterior edge of the embryo. The mid part of this becomes the notochord and the remainder the gut lining. The mesoderm becomes somites and lateral plate and the ectoderm becomes epidermis and neural plate, all arranged in the typical vertebrate body plan.

For the purposes of understanding the experimental work on the mouse embryo we do not need to describe its development any further except to note the derivation of the embryonic membranes at foetal stages. The amnion, as we have seen, is a covering of the dorsal surface of the embryo originally derived from ectoderm and mesoderm at the posterior end of the primitive streak. The visceral yolk sac is derived from the region of the egg cylinder composed of mesoderm and visceral endoderm which joins the ectoplacental region to the embryo proper. It enlarges considerably in later development and becomes apposed to the trophoblast/parietal endoderm layer which itself is sometimes known as the parietal yolk sac. The placenta is formed from the maternal tissues of the decidua, the ectoplacental cone, the chorion, and the allantois, which arises from the posterior mesoderm of the embryo.

It should be stressed that the above description holds only for the mouse. The course of development of mammals varies somewhat between species and in particular the arrangement of the extra-embryonic membranes can differ considerably.

Fate mapping

Fate maps in the usual sense do not exist for mammalian embryos since there is no satisfactory method of prospective marking. What information there is has actually come indirectly from studies of specification and determination. Once regions have become determined they must of course also have a predictable fate in normal development.

Škreb and his colleagues have studied the self-differentiation of explants from rat and mouse embryos transplanted to ectopic sites in the adult animal. They found that before gastrulation the isolated ectoderm could form tissues characteristic of all three germ layers (review by Škreb, Švajger & Levak-Švajger, 1976), which suggests that in normal development the definitive endoderm as well as the mesoderm is of gastrular origin.

Further studies have involved recombinations using glucose phosphate isomerase (GPI) isozymes as cell markers. GPI exists as two allelic forms, a and b, which have different electrophoretic mobilities on starch gels. Since different mouse strains are homozygous for different alleles it is possible to reconstruct embryos in which a particular region or cell type is of one variant and the remainder of the other. The embryos can then be implanted in foster mothers and allowed to grow. After a while they are recovered and dissected and each part of the embryo and its surrounding membranes examined for its proportions of the a and b isozymes. An advantage of this method is that GPI is an enzyme which is present in all tissues; but there are also some disadvantages. Methods involving embryo reimplantation are difficult and tedious, the spatial resolution obtained is only as good as the dissection, and it is not easy to detect proportions of one isozyme below a few per cent.

Gardner, Papaioannou & Barton (1973) and Papaioannou (1982) reconstructed blastocysts from ICM and trophectoderm of different GPI type, and were able to show that the ectoplacental cone, the secondary giant cells and the extra-embryonic ectoderm of later stages were derived from the trophectoderm. The entire foetus, the amnion, allantois and extra-embryonic endoderm were derived from the inner cell mass. When fragments or single cells from the primitive endoderm are injected into blastocysts they contribute to the placenta and to the visceral and parietal yolk sacs but not to the actual embryo of later stages (Gardner & Papaioannou, 1975; Gardner & Rossant, 1979; Gardner, 1982). In the visceral yolk sac it is the endoderm rather than the mesoderm that is labelled. This is the best evidence that all three germ layers of the embryo arise from the 'ectoderm' of the blastocyst and that the 'primitive endoderm' forms only extra-embryonic structures.

Cultured cells and teratocarcinoma

There have been several attempts to establish cultured cell lines from early mammalian embryos. Cole & Paul (1965) obtained strains of normal karyotype from 5–6-day rabbit blastocysts. Sherman (1975) obtained cultures from mouse blastocysts which were clonable, had a normal karyotype, and which developed into single-tissue tumours after implantation into syngeneic hosts. Recently Martin (1981) and Evans & Kaufman (1981) have produced pluripotent cell lines from ICMs and from egg cylinder ectoderm respectively. These will proliferate when grown on 'feeder layers' of irradiated cells and will differentiate into 'embryoid bodies' (see below) in the absence of feeder cells. When implanted into syngeneic mice they form tumours containing several differentiated tissue types. This behaviour is exactly like that of some of the cell lines derived from tumours called teratocarcinomas.

Teratocarcinomas are gonadal tumours which consist of several types of tissues and also contain undifferentiated cells. They are malignant and transplantable. The undifferentiated cells are often called 'stem cells' but in the present book this term is reserved for cells which undergo unequal divisions, and will not therefore be used, the alternative 'embryonal carcinoma' being preferred.

Three types of teratocarcinoma can be distinguished: spontaneous testicular, spontaneous ovarian and embryo-derived (Stevens, 1980). Spontaneous testicular teratocarcinomas arise in the testes of foetal male mice of strain 129. They are thought to arise from primordial germ cells and the well-known F9 cell line is of this type. Spontaneous ovarian teratocarcinomas arise in females of LT mice at about 3 months of age. They are derived from oocytes which have completed the first meiotic division (Eppig *et al.*, 1977) and undergo approximately normal embryonic development as far as the egg cylinder stage.

Embryo-derived teratocarcinomas can be produced by grafting early mouse embryos to extra-uterine sites in syngeneic hosts, usually the kidney capsule or testis. The embryos become disorganised and produce various adult tissue types together with proliferating undifferentiated cells. The latter seem to be derived from the embryonic ectoderm of 5–6-day embryos since they can arise from this tissue alone (Diwan & Stevens, 1976). A study of protein synthesis patterns on two-dimensional gels showed that embryonal carcinoma cells are more similar to 5-day egg cylinders than to ICMs from earlier stages or primordial germ cells from later stages (Evans, 1981).

Whatever their source, teratocarcinoma cell lines differ greatly in their properties one to another. Some will grow as ascites tumours in the peritoneal cavity and others will not. Some will differentiate *in vivo* or *in vitro* while others will not. Most lines have an abnormal karyotype although a few are normal or nearly normal. The type of differentiation which has been most intensively studied is the formation of 'embryoid bodies' which occurs in ascites tumours or *in vitro* (e.g. Martin & Evans, 1975). For the lines in question the cells remain undifferentiated while cultured on 'feeder layers' of inactivated cells but form embryoid bodies when plated out on their own. The bodies consist of an inner mass of embryonal carcinoma cells with an outer layer of cells somewhat resembling the primitive endoderm.

Teratocarcinoma enthusiasts argue that their cells can be regarded as model embryos on two main grounds: like early embryo cells they are multipotent, and although they arise as tumours they can be reprogrammed to normalcy by exposure to the appropriate environment. The multipotency is a single-cell property, as was first shown by Kleinsmith & Pierce (1964) who injected single cells into the peritoneal cavity, retransplanted the tumours which grew, and showed that a wide variety of differentiated cell types could be produced including occasionally types not formed by the parent tumour.

The reprogrammability was demonstrated by experiments in which teratocarcinoma cells were injected into the blastocoel of host embryos which could be distinguished by various genetic criteria such as coat colour and GPI isozyme, the host embryos being allowed to develop in foster mothers (Brinster, 1974; Mintz & Illmensee, 1975; Illmensee & Mintz, 1976). These authors used a testicular teratocarcinoma which grows as an ascites tumour in the form of embryoid bodies. Even when only a single cell was injected into the blastocyst a significant proportion of foetuses and full-term offspring could be shown to be composed partly of cells derived from the tumour. In subsequent experiments females with some oocytes derived from the tumour have been made and the production of functional gametes demonstrated by backcrossing (Illmensee, 1978). These experiments suggest that in at least some cases the teratocarcinoma cells can revert to normal behaviour and participate in the formation of a normal mouse. It is now hoped that new genetic strains of mouse can be produced by mutagenising and selecting the cells *in vitro* and then introducing them into embryos, producing germ line chimaeras and breeding homozygous mutant mice from the F_1 generation (Dewey & Mintz, 1980). The overall relationships between embryos, tumours and cell lines are shown in Fig. 6.2.

An important question which is not yet resolved and which the enthusiasts for this field have to answer concerns the diversity of teratocarcinoma lines. Does this in fact represent a wide range of

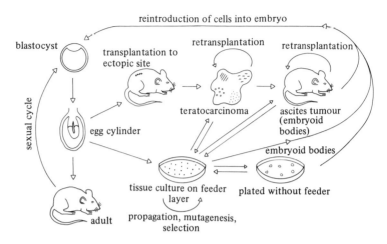

Fig. 6.2. Interrelationship between embryos, teratocarcinoma tumours and cell lines.

determined cell states to be found in the 5½-day embryonic ectoderm, or does it mean that most of the lines are abnormal and have departed to various extents from the *in vivo* precursor cell? The answer to this will presumably come in time from better biochemical characterisation of the regions within the embryonic ectoderm at early post-implantation stages. But in view of the abnormal karyotype of many teratocarcinomas it seems possible that cell lines grown directly from embryos may prove to be more normal and thus better models for working out the sequence of early developmental decisions.

Specification of early blastomeres

Since embryos of various amphibian, insect and sea urchin species will form twins when divided at early stages it should not come as any surprise that the same is true of early mammalian embryos. However, there is a difference. The forms dealt with earlier produce twins which are normally proportioned while mammalian blastocysts resulting from blastomere separation are not. This probably arises from the unique geometry of the first developmental decision ('inside *versus* outside'), which is in turn perhaps related to the fact that the early decisions are concerned with the specification of extra-embryonic parts rather than regions of the embryo itself.

The first blastomere separation experiment was carried out on the rat by Nicholas & Hall (1942). They implanted isolated blastomeres from the 2-cell stage ('½ blastomeres') into the uterus of pregnant rats and later recovered several embryos of the egg cylinder stage which they considered to be of a size intermediate between normal and half-normal. Seidel (1960) carried out an extensive series of experiments on rabbit embryos. He left the embryos inside the zona pellucida but destroyed one of the first two blastomeres or up to three of the blastomeres of the 4-cell stage. The embryos were implanted into pseudopregnant foster mothers of a different strain and allowed to develop for different periods of time. He found that most gave rise to normal embryos and that the size regulated to normal during primitive streak and neurula stages. Some cases were allowed to develop to term and to be born, and these later acquired a normal adult size and weight. Few abnormal embryos were formed although there was a tendency for the ¼ blastomeres to yield empty trophoblastic vesicles containing no embryo.

Similar experiments were carried out for the first time on the mouse by Tarkowski (1959). The results were similar and when the cell numbers in the blastocysts derived from ½ blastomeres were counted it was clear that the proportion of cells in the ICM was lower than normal. The development of isolated ¼ and ⅛ blastomeres *in vitro* was studied by Tarkowski & Wroblewska (1967). They showed that it was possible to obtain more than one blastocyst from the products of a single embryo but that these blastocysts, particularly from the 8-cell stage, frequently formed empty vesicles of trophectoderm. It therefore seemed as though all blastomeres had the same state of specification at early stages but the smaller the embryo the less likely it was to develop an ICM. This led to the idea that the first developmental decision occurred in the morula and that cells became specified to form ICM and trophectoderm depending on whether they lay in an inside or outside position.

Further evidence for the equivalence of early blastomeres came from experiments in which giant embryos were produced by the aggregation of two or more normal embryos at the 8- or 16-cell stage (Tarkowski, 1961; Mintz, 1965). These have a higher proportion than normal of cells in the ICM of the blastocyst and adjust to a normal size shortly after implantation (Buehr & McLaren, 1974). The aggregation of embryos of different strains produces a 'chimaeric' animal and these have been used a great deal in experimental work on developmental genetics (McLaren, 1976).

The most direct evidence for the importance of position in the establishment of the ICM and trophectoderm comes from experiments of Hillman, Sherman & Graham (1972). They labelled donor embryos with tritiated thymidine and then assembled morulae in which the labelled cells were either 'outside' or 'inside' a mass of unlabelled ones (Fig. 6.3). A labelled blastomere can be put outside simply by sticking it to another embryo at the 4- or 8-cell stage. In order to put it inside it has to be surrounded by several unlabelled embryos and thus eventually incorporated into a giant blastocyst. When the blastocysts were examined by autoradiography the

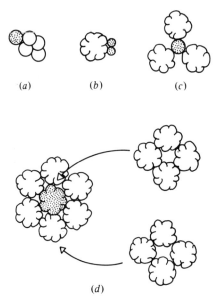

Fig. 6.3. 'Inside–outside' experiments of Hillman *et al.* (1972). In (*a*) and (*b*) early blastomeres from a labelled embryo are attached to the outer surface of another. These tend to become part of the trophectoderm. In (*c*) a labelled blastomere is surrounded by three unlabelled embryos, and in (*d*) an entire labelled embryo is surrounded by fourteen unlabelled embryos. In both these situations the labelled cells tend to contribute to the inner cell mass.

label was usually in the trophectoderm if the labelled blastomere had been outside, and in the ICM if the labelled blastomere had been inside. In the latter case it is possible to force an entire early embryo to become part of the ICM of a giant blastocyst. More recently all eight ⅛ blastomeres have been taken from a single embryo and shown to be able to contribute to ICM and trophectoderm after aggregation with eight separate host embryos (Kelly, 1977).

The exact way in which the difference of position is converted into a difference of cell state is not understood. However, it seems likely that cell surface interactions may be involved. If a giant blastocyst is formed by aggregating several embryos and then a single 8-cell embryo is injected without its zona it will adhere to one side and forms an 'ICM-like' mass, also contributing cells to the mural trophectoderm. However, if injected with a ruptured zona which allows access of blastocoelic fluid but does not allow the formation of cell contacts, then a normal blastocyst arises within the giant blastocyst (Pedersen & Spindle, 1980; Fig. 6.4). The blastocoelic fluid environment does not therefore seem to be a sufficient condition to persuade the cells that they are 'inside'.

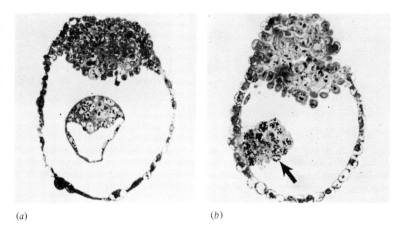

(a)   (b)

Fig. 6.4. Experiment in which an 8-cell embryo is injected into the blastocoel of a giant blastocyst, previously formed by the aggregation of several embryos. (a) With a ruptured or an intact zona the injected embryo forms a normal blastocyst; (b) without a zona it forms an ICM-like mass (arrow). (Photographs kindly provided by Dr R. A. Pedersen.)

In normal development it is known that microvillae arise on the outer surfaces of the blastomeres during compaction, and that the fourth cleavage is somewhat asymmetrical, yielding large outer cells with the villae and small inner cells without. The microvillous parts of the membrane can be visualised by the binding of fluorescein-conjugated concanavalin A (Handyside, 1980). It has been shown that ⅛ blastomeres *in vitro* can become polarised in the same way, especially if they are attached to another cell. They then usually divide at right angles to the polar axis to give the villous and avillous cell as in the intact embryo (Ziomek & Johnson, 1980; Johnson & Ziomek, 1981; Fig. 6.5). So the first step in the establishment of ICM and trophectoderm may be a self-polarisation of the blastomeres dependent on cell contact, and the second step an asymmetrical cell division. This is somewhat reminiscent of the situation in many invertebrate embryos discussed in Chapter 5, in which a cytoplasmic determinant becomes localised shortly before a stem cell division.

It should be stressed, however, that the commitment of the cells at this stage is not irreversible. Aggregates of all villous or all avillous cells of the ¹/₁₆ stage will still form complete blastocysts. Handyside (1978) and Hogan & Tilly (1978a, b) investigated the properties of inside cells isolated by immunosurgery, a technique in which an anti-mouse antibody is used to destroy the outer layer of cells. From morulae the inner cells will form complete blastocysts, from early blastocysts they will sometimes do so,

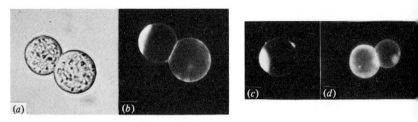

Fig. 6.5. Polarisation of mouse embryo blastomeres. (*a*) and (*b*) show two ⅛ blastomeres which are derived from an isolated ¼ blastomere. The cell has divided and a membrane polarisation has developed visualised with the fluorescent label FITC–ConA. (*a*: phase-contrast image; *b*: fluorescence image). The polarisation is more apparent for the left-hand cell. (*c*) and (*d*) show that when an isolated ⅛ blastomere divides it can generate dissimilar daughters as visualised by FITC–ConA binding. (Photographs kindly provided by Dr M. Johnson.)

and from later blastocysts they never do so. By this stage the two tissues appear to have quite different properties. The trophectoderm will accumulate fluid, become implanted in the uterus and stimulate a decidual response, but produces only a few giant cells and no embryonic structures (Gardner, 1972). The ICM by contrast will not implant in the uterus. *In vitro* it will readily fuse with other ICMs, and develops an outer layer of primitive endoderm. In addition, the two tissues have significantly different patterns of protein synthesis as visualised by two-dimensional electrophoresis (Van Blerkom *et al.*, 1976), and, as will be discussed later, in female embryos one X chromosome is inactivated in the trophectoderm but not the ICM (Monk & Harper, 1979).

Recently nuclear transplantation studies have been carried out. The nucleus from an ICM or a trophectoderm cell was injected into recently fertilised eggs and the host pronuclei removed with the same micropipette (Illmensee & Hoppe, 1981). The ICM nuclei were significantly better at supporting development to the blastocyst stage than the trophectoderm nuclei and three mice were carried to term out of sixteen reimplanted blastocysts. So as far as ICM nuclei are concerned the situation resembles that in Amphibia or *Drosophila* in that they will support normal development of the egg in a small proportion of cases. It is possible that trophectoderm nuclei are 'more differentiated' than ICM nuclei since they do not give such good results, but it is hard to assess negative results in this type of experiment.

Later developmental decisions

Shortly before the formation of the endodermal layer it is possible to distinguish cells of the ICM as 'smooth' or 'rough' on the basis of their appearance after disaggregation. These cells can be assayed for their state of determination by injecting them into blastocysts of different GPI isozyme type, implanting into foster mothers, and analysing the later conceptuses for isozymic composition (Gardner & Rossant, 1979). The smooth cells were found to contribute to the embryo, the amnion, the mesoderm of the visceral yolk sac and the placenta. The rough cells contributed to the endoderm of the yolk sac and to the placenta. These differences probably imply that the primitive endoderm becomes determined shortly before its appearance as a visible cell layer.

It is thought probable that the initial cue for this determination is the position of the cells within the ICM. This is because isolated ICM aggregates develop a layer of endoderm all around the outside, as do teratocarcinoma cells when they form embryoid bodies. Also reduction of the number of ICM cells can lead to conceptuses in which primitive endoderm derivatives are present but in which there is no embryo (Gardner, 1978). In this case, as with the early embryo, there seems to be little or no regulation of proportions in the short term although adjustment can occur later so long as some cells of each population are present.

An environmental difference also probably underlies the subdivision of the trophectoderm into mural and polar regions, the polar tissue forming above the ICM. There are two pieces of evidence for this. Firstly, trophectoderm which develops without an ICM, such as that from ¼ or ⅛ blastomeres, is mural all over. Secondly, when an extra ICM is introduced, as in the rat–mouse chimaera experiments of Gardner & Johnson (1973), a polar region forms over the second ICM as well as over the first.

Little is known about regional subdivisions within the ectoderm but it seems likely that environmental differences also underlie the formation of visceral and parietal endoderm from the primitive endoderm. The visceral endoderm is characterised by synthesis of α-foetoprotein, transferrin and fibronectin while parietal endoderm makes type IV collagen and laminin but no fibronectin. Hogan & Tilly (1981) found that the visceral endoderm would maintain its properties when cultured in isolation but in the presence of extra-embryonic ectoderm would become parietal endoderm. An analogous situation exists for F9 teratocarcinoma cells. These can be induced to form a primitive endoderm-like tissue by treatment with retinoic acid (Strickland & Mahdavi, 1978). If this is then treated with cyclic AMP it becomes parietal while if it is recombined with embryonal carcinoma cells it becomes visceral (Hogan, Taylor & Adamson, 1981). These experiments represent an interesting combination of studies using embryos and teratocarcinoma cells, an approach we shall probably see more of as procedures for inducing differentiation in teratocarcinomas improve.

Chimaerism and mosaicism

We have already encountered genetic mosaics in Chapter 4. Among mammals an animal composed of cells which are genetically dissimilar is called a mosaic if it has grown that way, and a chimaera if it has arisen from some experimental mixture of cells or embryos (McLaren, 1976). Mammalian mosaics and chimaeras have been used extensively in recent years to investigate problems in developmental genetics and in late development, and some of the results also have a bearing on early events.

Mintz (1970) carried out a series of aggregations of morulae from two strains of mice. About 75% of the resulting mice were chimaeric and 25% non-chimaeric. This was interpreted to mean that about three cells gave rise to the embryo and the remainder to extra-embryonic structures. The reason is that if the cells are randomly mixed in the aggregate and a region consisting of n cells is fated to become the embryo then the probability that an embryo will be drawn from cells of one type is $2 \times (\frac{1}{2})^n$, which is 0.25 for $n = 3$. This estimate of the proportion of the morula which becomes the embryo should not, however, be regarded as a very accurate one because a cell will be counted if any of its progeny enters the embryo (McLaren, 1972), and because cells are not randomly mixed in pre-implantation aggregation chimaeras anyway (Garner & McLaren, 1974).

All female mammals are naturally mosaic for X-linked heterozygous loci because of the inactivation of one X chromosome in every cell by heterochromatinisation. It is now known that X-inactivation occurs in different tissues at different times and in particular that it occurs in each of the extra-embryonic parts shortly after their determination, the paternal X chromosome being preferentially inactivated (Monk & Harper, 1979). Inactivation occurs in the embryo itself at 6–6½ days, which is the stage of the primitive streak and, presumably, of determination of the embryo body plan. It is apparently random with respect to the maternal or paternal origin of the chromosome. X-inactivation is one of the features, together with ultrastructure and early synthesis of terminal differentiation products, which indicates a precocious development of the extra-embryonic parts relative to the embryo itself.

In view of the lack of cell mixing in pre-implantation chimaeras in which one component is labelled with tritiated thymidine (Hillman *et al.*, 1972; Garner & McLaren, 1974), it is surprising that late embryos and adult female mice show an extremely fine-grained mosaicism with respect to X-linked genes for which they are heterozygous, having, for example, about 9000 coherent clones per retinal epithelium in the eye (Deol & Whitten, 1972; West, 1976). The same is true of aggregation chimaeras. The cell number in the mouse embryo at 6½ days is about 800 (Snow, 1976), so if clones remained coherent after this time there should be somewhat less than 800 patches in the whole animal. In fact there are many more and this is presumably due to the breaking up of clones during gastrulation and perhaps again during organ formation when many organs are formed by extensive convolutions of neighbouring tissue layers.

Twinning

We have seen that twins can be produced experimentally by separation of the first two blastomeres. But twinning has also been observed to occur by subdivision of the ICM in embryos cultured *in vitro* (Hsu & Gonda, 1980) and can be induced by injecting the mothers with the cytotoxic drug vincristine on day 7 (Kaufman & O'Shea, 1978). This implies that twins can arise from disruption of the embryo as late as the head process stage. Human monozygotic twins usually arise from division of the ICM (70–75%), less often from blastomere separation (25–30%) and most infrequently from division of the primitive streak (1%). These figures are arrived at on the basis of whether the twins share a common placenta and amnion (Hamilton & Mossman, 1976). The possibility of twinning in the gastrula indicates that, as in amphibians, the specification of parts in the transverse axis of the body is still labile at this stage.

Comparison with other types of animal

In conclusion, we see that the determination of the extra-embryonic parts in the pre-implantation mouse embryo bears certain similarities to early decisions in other embryos. Regional distinctions are set up in small groups of cells in response to environmental cues. The distinct states are initially labile (specification) but soon become stable and clonally heritable (determination). What has been shown clearly for the first time by the mouse experiments is that single cells can become determined. Determination is not therefore a property only of grafts containing many cells. In the literature a certain amount of stress has been laid on the fact that the decisions are all binary ones, since binary decisions are sometimes thought to have an intimate relationship with cell division (Holtzer, 1978). However, the spectacular creation of giant embryos by aggregation of morulae, and their size regulation shortly after implantation, argue against any rigid connection between the steps of determination and the cell cycle.

There are two unusual features of determination in the mouse embryo. First, there is not an exact control of proportions following experimentally induced size changes, either with respect to the ICM:trophectoderm or with respect to the primary ectoderm:endoderm ratios. Secondly, there seems to be a precocious terminal differentiation of one component after each binary decision, as assessed by synthesis of major differentiation products, X-inactivation and, perhaps, the inability of transplanted nuclei to support development of the egg.

It is possible that these two features are associated. The extra-embryonic membranes are called upon to function soon after implantation and so early differentiation is necessary in a way which it is not for the embryo itself. So perhaps the stage of labile specification for these components is very short compared with that of the ICM and ectoderm respectively, and there is insufficient time for regulation of proportions.

Alternatively it may be that the mechanisms producing the regional subdivision of embryonic from extra-embryonic structures bear no relation to those involved in subdivision of the embryo itself, and are intrinsically incapable of proportion regulation. This view is supported by the lack of evidence for any localised organiser regions which have been a fairly general feature of the inductive interactions considered in the other chapters. However, those readers with a general commitment to universality of mechanism will find it an unpalatable conclusion.

The chick embryo

The early chick embryo resembles the mouse in several respects. The early blastoderm becomes divided into two layers somewhat like the ICM, and the gastrulation movements occur through a primitive streak which forms at the posterior end. The embryo undergoes extensive growth during development and devotes a substantial proportion of its early tissue to extra-embryonic structures, some of which – the amnion, chorion and allantois – are clearly homologous to those of mammals.

For the experimentalist the chick has the great advantage over the mouse or rabbit that the embryo is accessible at all stages. Early blastoderms can be cultured *in vitro* for long enough to form a recognisable primary body plan (New, 1966), or can sometimes be manipulated *in ovo* and kept alive until later stages. Furthermore it is possible to explant small pieces of tissue onto the chorioallantoic membrane of later embryos, where they become vascularised and will grow and differentiate in isolation. For these reasons our knowledge of the normal morphogenetic movements, the fate maps and the early determinative events in the chick is far superior to our knowledge of the post-implantation stages of mammalian development.

The studies of inductive interactions tend to show that the situation is generally similar to that in amphibian embryos. This is most interesting because superficially amphibians are very different from the higher vertebrates in size, structure and gastrulation movements. Of course the general vertebrate body plan which is achieved by the end of early development is similar in Amphibia, birds and mammals, and if the method of getting there in terms of the hierarchy of determinative decisions is also similar then we can probably conclude that the visible differences are superficial ones, presumably adaptations to the different modes of embryonic life of the animals concerned.

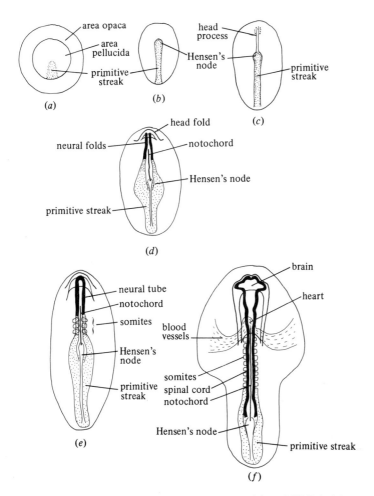

Fig. 6.6. Normal development of the chick embryo. (*a*) and (*b*) Primitive streak stages, (*c*) head process, (*d*) neurulation, (*e*) somite formation, (*f*) early tailbud.

Normal development

At the time of oocyte maturation the future embryo consists of a germinal vesicle and small patch of cytoplasm at one edge of an enormous yolk mass. Fertilisation occurs in the oviduct and by the time of laying the embryo has become a blastoderm of about 60 000 cells, of which only about 500 will contribute to the embryo proper (Spratt & Haas, 1960).

The blastoderm is initially thicker in the centre but the cells rearrange themselves so that a thick area opaca is formed at the periphery and a thinner area pellucida in the middle (Fig. 6.6). The area pellucida is two

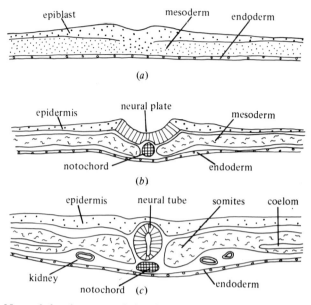

Fig. 6.7. Normal development of the chick: transverse sections. (*a*) Primitive streak, (*b*) neural plate stage, (*c*) neural tube stage.

or three cells thick except at the future posterior end where it may be five cells thick. A lower layer, called the hypoblast, develops, partly by 'polyinvagination' and partly from the posterior area opaca (Vakaet, 1962; Spratt & Haas, 1965). This tissue contributes only to extra-embryonic structures and may perhaps be homologous to the primitive endoderm of mammals.

A condensation of cells known as the primitive streak arises from the posterior of the area pellucida and elongates until it reaches the centre. Cells in the epiblast, or upper layer, migrate laterally into the streak and pass through it to become the mesoderm and the endodermal part of the hypoblast. These movements have been followed using orthotopic grafts labelled with tritiated thymidine (Nicolet, 1971) and it now seems quite clear that the embryonic endoderm as well as the mesoderm is of gastrular origin.

The area pellucida gradually changes from a disc to a pear shape and a condensation called Hensen's node appears at the anterior end of the primitive streak. This contains the presumptive notochord cells which migrate anteriorly to form the head process. The remainder of the node moves posteriorly and as it does so the principal structures of the body plan appear in its wake (Figs. 6.6 and 6.7). By about 1 day of incubation the anterior end of the embryo is marked by a wave in the blastoderm called the head fold, and one somite and the anterior neural folds have appeared in the track of the regressing node. By about 36 hours there are ten somites and the neural tube has closed to form three brain vesicles. Although node regression and the formation of the posterior part of the body continue for some time, this stage marks approximately the junction between early and late development since the general body plan has been laid down and the formation of individual organs is about to begin.

Fate maps and isolation experiments

The prospective regions in the chick blastoderm have been studied for almost as long as those of the amphibian but the fate maps are not as accurate, perhaps because of the small size of the early blastoderm or perhaps because there is more random mixing of cells and so the fate maps have an inherently lower resolution.

The three techniques which have been used for marking are vital staining, carbon particles and orthotopic grafts labelled with tritiated thymidine. Each worker has disagreed to some extent with the others and so in Fig. 6.8 are shown 'compromise' fate maps devised by Waddington (1952) for the early blastoderm and a vital staining experiment of Nicolet (1970) on the head process stage. One notable aspect is the extent to which the node region itself is the prospective region for the axial structures. The fact that the posterior part of the streak will not form notochord, somites or neural tube in isolation is not in the least surprising since its fate in normal development is to become extra-embryonic mesoderm. It is also interesting that structures which later on are arranged mediolaterally are derived from presumptive regions which at the head process stage are arranged concentrically around the node.

Several workers have cut chick blastoderms into fragments and cultured them on the chorioallantoic membrane of late embryos. This is regarded as a 'neutral site' from the point of view of induction and so the procedure is a test for specification similar to the culture of explants from amphibian embryos in buffered salt solution, or of invertebrate blastomeres in sea water. In Fig. 6.9 are shown some results of Rawles (1936) for culture of fragments from head process stage embryos. Unlike the original diagrams this one depicts only significant proportions of positive cases and it is really very similar to Nicolet's fate map, thus showing essentially mosaic behaviour. This suggests that by the head process stage the decisions involved in the specification of the major structures of the body have already taken place. Some isolation experiments have been carried out on earlier stages but unfortunately involved cutting the blastoderm into only two or three pieces, which does not provide sufficient resolution for detecting departures from mosaic behaviour.

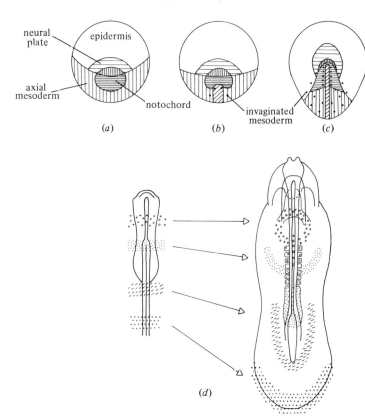

Fig. 6.8. (*a*)–(*c*). Fate maps of the chick epiblast during establishment and elongation of the primitive streak. (After Waddington, 1952.) (*d*) A marking experiment by Nicolet (1970) on the head process stage showing that the posterior part of the primitive streak does not normally contribute to the body axis.

Regulation of the blastoderm

An extensive study involving fragmentation of the blastoderm *in ovo* was conducted by Lutz (1949) using the duck. In this species the egg is laid at a slightly earlier stage than in the chicken. He found that up to four complete embryos could be produced by cutting (Fig. 6.10). Spratt & Haas (1960, 1961) obtained similar results with the chick, showing that an embryo could arise from as little as an eighth of a blastoderm or as much as three blastoderms fused together. The size of the embryos themselves did not vary by a factor of 24, but more like a factor of 3 in volume. This is another indication that the proportions within the embryo are accurately controlled but the proportion of tissue devoted to extra-embryonic as opposed to embryonic tissues is not.

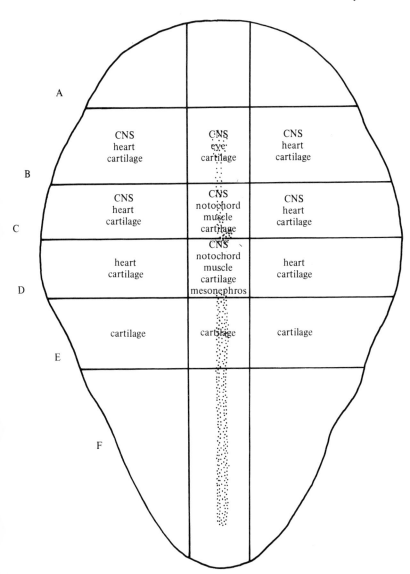

Fig. 6.9. Tissues formed from isolates of head process stage embryos grown on the chorioallantoic membrane. Only cell types formed in a high proportion of cases are shown.

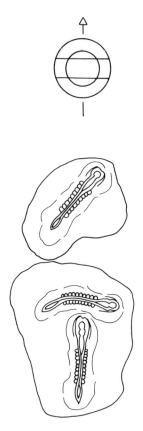

Fig. 6.10. Formation of triplets from a single duck blastoderm cut into three pieces. The posterior embryo has the original orientation but the anterior and middle embryos do not. After Lutz (1949).

The future position of the primitive streak is already determined by the time of laying and finds expression in the rule of von Baer according to which the posterior lies towards the investigator if he holds the egg with the pointed end to the right. According to Clavert (1963) this is only true if the egg is laid pointed end first and the implication is that the anteroposterior axis is determined while the egg is in the uterus. Lutz showed that when the blastoderm was cut parallel to the future axis both of the twins would be of normal orientation, but if the cut was perpendicular to the axis only the posterior twin would be normally oriented and the anterior one would be randomly oriented. For triplets, only the most posterior is normally oriented (Fig. 6.10). This suggests that the situation resembles that of the amphibian dorsoventral axis. The system is poised to become polarised in any orientation but a slight bias will make the process deterministic. If the polarisation commenced with, say, a slight accumulation of cells at the posterior, then posterior fragments would retain their polarity while anterior ones would become repolarised by any slight environmental heterogeneity to which they are exposed. The theory of symmetry-breaking processes will be discussed in Chapter 8.

Induction

The situation regarding induction resembles that of the amphibian organiser in the sense that the literature tends to refer to a single inductor or organiser when there are in fact several interactions taking place in early development.

The earlier work, reviewed by Waddington (1952), suffers from the lack of a reliable cell-autonomous marker which can distinguish between donor and host tissues. Without such a marker it can be difficult to decide which structures have really been induced from the host and which are formed by self-differentiation of the graft. In recent years two markers have been introduced and the situation has accordingly become a little clearer. The first is tritiated thymidine, which has already been referred to in connection with fate mapping. The second involves interspecies grafts between the chick and the quail. Cells of the quail were shown by Le Douarin (1971) to contain a heterochromatic blob in the interphase nucleus which could be stained with the Feulgen reaction to provide an unambiguous positive marker. The quail marker is not suitable for fate mapping since the embryos of the two species differ somewhat in developmental rate, but it is quite acceptable for experiments on induction and in some respects it is surprising that it has not been more widely used.

We have seen that the primitive streak normally originates at the posterior end of the area pellucida epiblast. There is some evidence that a second streak can be induced elsewhere by the presence of the posterior hypoblast. Thus Waddington (1933) using both ducks and chicks obtained three cases of embryos with two streaks by rotating the hypoblast through 180 degrees. Gallera & Nicolet (1969) were able to induce primitive streaks from the area opaca by implanting the middle region of a streak labelled with tritiated thymidine, the grafts themselves differentiating into endoderm. These two experiments are probably equivalent and suggest that the position and polarity of the blastopore is established by an interaction between the hypoblast and epiblast. Azar & Eyal Giladi (1979) showed that hypoblasts regenerated by isolated epiblasts were active as inductors but only if they contained cells from the region round the edge where the area pellucida borders on the area opaca. The effect bears an obvious parallel to the experiments of Nieuwkoop which demonstrate that in Amphibia the dorsoventral axis of the mesoderm, and hence the position of the blastopore, is determined by the endoderm (see Chapter 3).

The next possible interaction is a regionalisation of the mesoderm under the influence of the node. Bellairs (1963) showed that at the head process stage posterior-third blastoderms would not form somites unless a thin strip of anterior streak material was included. Nicolet (1970) grafted nodes labelled with tritiated thymidine to such posterior isolates and found that somites were formed from host tissue while the grafts differentiated mainly into notochord. Hornbruch, Summerbell & Wolpert (1979) grafted quail nodes at varying distances from the host streak of complete head process stage chick blastoderms. They found that when the graft was far away it self-differentiated into notochord, somites and some neural tissue, but when it was close to the host streak then extra somites could be induced from host tissue.

These experiments suggest that the mesoderm can be raised from the level of specified extra-embryonic mesoderm to specified somite under the influence of the node, the prospective regions for these tissues being arranged in concentric rings around the node. This bears an obvious similarity to the dorsalising activity of the amphibian organiser discussed in Chapter 3 and a 'gradient' model for the process was advanced by Hornbruch *et al.* (1979). It is worth noting that both the node and the early dorsal lip will often re-form if they are cut out. This behaviour is not typical of other organisers but does pose a problem for the model-builders and will be discussed further in Chapter 9.

The best characterised inductive interaction in the avian, as in the amphibian, embryo is neural induction. Waddington (1932) reported that neuroepithelium could be induced from the epiblast of the area pellucida and area opaca in response to implants of primitive streak. He even obtained two cases of induction from the embryonic shields of rabbit embryos after implanting chick primitive streak (Waddington, 1934). These experiments were, however, performed without cell markers. Grabowski (1957) repeated the experiments with vitally stained nodes from definitive streak stages and found that the neural tissue was indeed derived from the host. More recently the interaction has been investigated in detail by Gallera (1971) using grafts labelled with tritiated thymidine. Fig. 6.11 shows an experiment in which a quail node has been grafted into a chick area pellucida and induced a neural plate from the host.

Inductive activity initially lies in the anterior primitive streak and later in the region just anterior to the regressing node. It disappears by about the four-somite stage. The competence of the ectoderm disappears before this, at the head process stage. About 4 hours of contact are required between the inductor and the area pellucida epiblast in order to obtain a response, and the signal will pass across a Millipore filter with a nominal pore size of 0.45 μm. Unlike neural induction in Amphibia, the

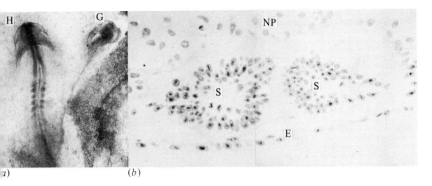

Fig. 6.11. Graft of Hensen's node from a quail to a chick blastoderm. (*a*) The secondary axis formed from the graft. G, graft; H, host. (*b*) A section through the secondary axis. The somites (S) and endoderm (E) are graft-derived and the neural plate (NP) is host-derived. (Photographs kindly provided by Dr A. Hornbruch.)

effect is abolished if the inductor is killed by boiling or freezing, and it is not easily provoked by foreign tissues or substances.

Since the three types of induction which can be distinguished in the chick have obvious homologues in the Amphibia it seems probable that both will turn out to be valid models for the mammals in the sense that the biochemical basis of the interactions will be the same, even though the morphogenetic movements of the early stages appear to be substantially different. Homologies with the invertebrate embryos are, however, not so obvious. As we shall see in the next chapter there are several phenomenological resemblances but it is clearly not possible to find processes which are 'the same' as the early inductive interactions in the sense in which we understand them in the vertebrates.

General references

Waddington, C. H. (1952). *The Epigenetics of Birds*. Cambridge University Press.
Bellairs, R. (1971). *Developmental Processes in Higher Vertebrates*. Logos Press, London.
Embryogenesis in Mammals (1976). CIBA Foundation Symposium 40. Elsevier, Amsterdam.

Part III

Theories and models

7

The problems of early development
and the means for their solution

The preceding four chapters have provided a rapid trip through what is known about early cellular and regional commitment in a variety of types of embryo. It may be helpful at this stage to attempt to summarise the salient results.

In Table 7.1 the types of embryo we have considered are classified with respect to seven general properties. This reveals a broad measure of similarity between their behaviours which may earlier have been obscured by the mass of experimental detail. The similarities become even more apparent when we examine the 'no' entries in the table more carefully.

Properties 1 and 2 concern regulation of proportions in response to a change in the overall size of the embryo. All the embryos can regulate their proportions except the early mouse, and as has been discussed previously this may be because there is something special about the segregation of extra-embryonic as opposed to embryonic parts.

Property 3 concerns cytoplasmic localisation, the test being whether any rearrangement of cytoplasm can cause an alteration in the later pattern. The most common experiment involved here is low-speed centrifugation, which sometimes does and sometimes does not have an effect. However, in the cases where it does (certain amphibians and insects) the regime required is quite precisely defined and so it is possible that in the cases where it has no effect an appropriate regime has not yet been found. The other 'yes' entries refer to compression experiments which alter the early cleavage planes. These do not affect amphibians but can affect annelids, ascidians and molluscs.

Property 4 refers to the determination of isolated nuclei. In the cases examined it seems likely that there is none up to the blastula stage or its equivalent. It is, however, difficult to assess the negative results – for example from the mouse trophectoderm, or from post-gastrulation stages of insects and Amphibia – since an inability to support development is not a good criterion for determination.

Properties 5 and 6 refer to the stage of first specification and to the possibility of twinning. These are clearly associated since the embryos with early specification do not form twins and those with later specification

165

Table 7.1. *Classification of embryos with respect to seven general properties*

Property	Mammal	Bird	Amphibian	Sea urchin	Insect	Ascidian	Annelid	Mollusc	Ctenophore
1	No	Yes	Yes	Yes	Yes	Yes	Yes	Yes	Yes
2	No	Yes	Yes	Yes	?	Yes	?	?	?
3	?	?	Yes	No	Yes	Yes	Yes	Yes	No
4	Yes	?	Yes	?	Yes	?	?	?	?
5	32-cell	?	64-cell	32-cell	?	4-cell	2-cell	2-cell	8-cell
6	Yes	Yes	Yes	Yes	Yes	No	No	No	No
7	Yes	Yes	Yes	Yes	Yes	Yes	?	Yes	?

1. Adaptation of proportions to reduction in size.
2. Adaptation of proportions to increase in size.
3. Is there any known redistribution of cytoplasm which can affect the pattern of the embryo?
4. Totipotency of early nuclei.
5. Approximate cell number at which specification of parts can be detected by isolation experiments.
6. Can complete twinning be induced by division of the early embryo?
7. Evidence for intercellular interactions involved in regional specification.

do form twins. This difference corresponds to the distinction between 'regulative' and 'mosaic' embryos made in the early part of this century. Nowadays it seems less likely that this implies a difference in the underlying mechanism. We know that both cytoplasmic localisation and inductive interactions are widespread, if not universal. We also know that there are significant variations within the animal groups shown in the table. For example, some insect embryos form twins on longitudinal constriction while others, such as *Drosophila*, do not, and it seems unreasonable for radically different mechanisms to exist for organisms whose descriptive embryology is so similar.

In Tables 7.2 and 7.3 are listed the examples of cytoplasmic localisation and of induction which have been mentioned in the previous chapters. These are *prima facie* rather than proven cases, the amount of evidence varying from very good to a single believable experiment. An invariant cell lineage or the occurrence of mosaic behaviour following blastomere separation is not in itself considered to be evidence for cytoplasmic localisation. The examples in Table 7.2 are of cases in which a change in the cytoarchitecture has produced a change in the later pattern. In no case is it really proved that the 'determinant' actually consists of molecules which can regulate gene expression: the effects may be a good deal less direct than this. The examples of induction in Table 7.3 are probably all instructive rather than permissive but this cannot be known with confidence in the cases where there is a long time delay between the signal and the observation of its effects. For both cytoplasmic localisation and induction an attempt has been made in these tables to distinguish between the immediate response and the long-term consequences. The critics of induction are quite right in insisting that it is not possible to induce neurons from cleavage blastomeres, rather it is early decisions in the developmental hierarchy which are affected and the consequent later cytodifferentiation which we observe.

It should be clear by now that quite a lot is known about early embryonic development, and much of what is not known will eventually become known as a result of pursuing existing lines of research. It seems that a reasonably satisfactory account at the cellular level needs to comprise: (1) a fate map of the highest possible resolution, (2) a map of labile and stable cell states at each stage of development, (3) the times and places of cytoplasmic localisation and the orientation of stem cell divisions, (4) enumeration of organisers, the timing and range of their signals, and the competence of the responding tissues, and (5) rules for the transformation of one pattern of cell states into another by cellular and tissue movements.

At present the best-understood embryos are those of the Amphibia, for which some information exists for points 1, 2 and 4, but none for 3 and 5. Clearly a great deal of research remains to be done, and if one is

Table 7.2. *Cytoplasmic localisation*

Animal	Stem cell	Visible marker associated with 'determinant'	Stages of localisation	Parts specified	Structures formed subsequently
Insect	Egg	Pole plasm	Oogenesis	Pole cells	Germ cells, part of mid-gut
Mollusc	Egg	Polar lobe	Oogenesis or fertilised egg	D blastomere	Somatoblast, mesentoblast, organiser?
Annelid	Egg	Polar lobe	Fertilised egg	D blastomere	As above?
Ascidian	½ Blastomere	Yellow crescent	Fertilised egg	Posterior parts	Muscle, mesenchyme
Amphibian	½ Blastomere	Grey crescent	Fertilised egg	Dorsal parts	Notochord, CNS
Ctenophore	¼ Blastomere	None[a]	4-cell	e micromeres	Comb plates

[a] 'Ctenoplasm' in *Beroe* segregates to all micromeres, whereas the 'comb plate determinant' only segregates to e micromeres.

Table 7.3. *Inductive interactions*

Animal	Organiser region	Region which becomes subdivided	Stage of interaction	Markers whose later appearance reveals interaction
Mouse	None?	Whole embryo	Morula	ICM + trophectoderm
	None?	ICM	Blastocyst	Primitive endoderm
	ICM	Trophectoderm	Blastocyst	Polar trophectoderm
Chick	Hypoblast	Epiblast	Blastoderm	Primitive streak
	Anterior primitive streak	Mesoderm	Primitive streak	Somites
	Anterior primitive streak	Ectoderm	Primitive streak	Neural tube
Amphibian	Vegetal hemisphere	Animal hemisphere	Blastula	Mesodermal tissues, blastopore
	Dorsal mesoderm	Mesoderm	Blastula	Somites, kidney, blood
	Archenteron roof	Ectoderm	Gastrula	Neural tube
Sea urchin	Vegetal hemisphere	Animal hemisphere	Morula	Proportions of stereocilia and gut in prism stage
Insect	Posterior pole	Cleavage syncytium	Cleavage	Sequence of segments
Mollusc	D macromere	Animal micromeres	Cleavage	Eyes and statocysts
Ascidian	A4.1 blastomeres or progeny	a4.2 blastomeres or progeny	Unknown	Neural tube

principally interested in medical applications then the work has hardly started, since our knowledge of human embryology remains confined to an account based on descriptive histology. However, the main point is that much of this type of cellular level explanation is not reducible to lower levels, and it is certainly never going to be possible to understand how an embryo develops without a more satisfactory explanation at the cellular level.

In some respects the situation resembles that in human or animal physiology, subjects often neglected by academic molecular and cell biologists. Our understanding of the heart, for example, is mainly in terms of macroscopic variables. We consider the control of its output by filling pressure and by sympathetic stimulation. We consider the effects on the circulation of venous capacity, elasticity of the arteries and constriction of arterioles brought about by nervous and hormonal stimulation. Of course all the parts of the cardiovascular system also have biochemical problems to offer – how the cardiac myofibrils work, how the hormone receptors work, and so on – but it is clear that our understanding of how the system works as a whole is primarily electrical and mechanical and is going to stay that way.

Up to a point our understanding of embryos will likewise always remain at the cellular level. However, the situations are not quite comparable because in embryology we think that there are certain questions whose answers we expect in biochemical language which bear not just on the detailed operation of the components of a system but on the basic properties summarised in the tables. These are: firstly, the generation and maintenance of an inhomogeneous prepattern; secondly, the ability of this prepattern to regulate its proportions; thirdly, the conversion of the prepattern into stable and clonally heritable cellular states; and fourthly, why these states are associated with particular types of cell movement and terminal differentiation.

The last point lies outside the scope of this book and will not be discussed; the first three comprise the subject matter of the next three chapters. The term 'prepattern' is here used in a rather general sense to mean the distribution of any morphogenetically active substances whether they are involved in cytoplasmic localisation or induction.

It should further be noticed that all of these mechanistic problems themselves exist at two levels. At the general level we want to know what types of mechanism are capable of producing the observed behaviours, and at the particular level we want to know the identities of the substances involved and the details of their kinetic interrelationships. The relative importance of these two levels will only be known after the event. If it turns out that the biochemical substances involved are very numerous, or are different from one type of embryo to the next, then it will clearly be the general type of mechanism that is of most interest, although it will still be necessary to work out the human embryo in detail for medical reasons. On the other hand if it turns out that the biochemical mechanisms are simple and universal then they will be regarded as more important than possible mechanisms which may exist on paper but not in reality.

To reiterate: *the cellular level* of explanation is essential in its own right and not simply as a statement of problems for further reduction. We can anticipate greater understanding at this level by extension of the type of knowledge which we already possess.

The *general level of mechanism* is a theoretical study of the constraints which observed properties place on possible mechanisms. In the next three chapters we shall investigate this area of theoretical biology more closely. We shall find that the results of theoretical work have succeeded in demystifying the problems but are unable to prove the existence of specific mechanisms, far less suggest the identities of the substances involved.

The *particular level of mechanism* names the substances and specifies their quantitative behaviour. Nothing whatsoever is known about this at the moment and it is difficult to be confident in prescribing a line of research. We know of very many substances which are present in cells and there are doubtless many more remaining to be discovered. A search for key substances involved in regional specification resembles the search for a contact lens in a swimming pool, with the added uncertainty that the lens may have dissolved in the water.

8
Cell states

The term 'cell state' has been used repeatedly in connection with specification, determination and differentiation but has not yet been defined. In this chapter we shall investigate more closely what is meant by a cell state and consider the implications of this for the understanding of the biochemical basis of the early embryonic phenomena mentioned in the previous chapter.

Known properties of cell states

Most of what we think we know about cell states is derived from the observed properties of the histologically distinguishable cell types in the adult animal. In the human there are about 210 cell types outside the central nervous system (J. Lewis, personal communication). We do not know whether one cell *type* corresponds to one or many cell *states* since the histological appearance of cell types is dominated either by shape or by massive accumulations of rather small numbers of proteins such as haemoglobin in the red blood cell or actomyosin in striated muscle. Biochemical differences in components of low abundance are not of course visible down the microscope. If the principle of non-equivalence discussed in Chapter 1 is correct, then a rather generally occurring cell type such as smooth muscle would actually correspond to a different cell state in each part of the body in which it was found.

The cell types that we can see are notable in three respects. They are, on the whole, qualitatively distinct from one another without intermediate forms. They persist for long periods of time: in the case of myotubes and neurons for the entire life of the animal, and in other cases for weeks or months. Those tissues which are capable of proliferation ('the expanding compartment': Leblond, 1972) maintain their cell type when they divide. These generalisations are illuminated to some extent by their exceptions. Where it is possible to find a grading of one cell type into another this is usually a sequence of maturation, for example in the epidermis or the bone marrow. So we associate continuity of type with progressive change and we associate discreteness of type with stability.

Table 8.1. *Characteristics of differentiated and embryonic cells*

Terminally differentiated cells	Embryonic cells (and cells involved in adult tissue renewal)
1. Qualitatively distinct types	Types grade into one another in space and time
2. Stable for long periods	Spontaneously change in time
3. Retain type on cell division	May change on cell division
4. Rarely change type and need prior dedifferentiation	Easy to divert from one pathway to another

The same is true in the embryo: the early stages show no histological differentiation at all and when regions first become visible the boundaries are initially vague and indistinct and only later become sharp and clear-cut.

On the rare occasions when cells do seem to change their visible type, this is invariably an accompaniment of regeneration and involves prior dedifferentiation. Amphibian limb dermal fibroblasts can become cartilage (Dunis & Namenwirth, 1977), pigmented retina can become lens (Yamada, 1967), and in crustaceans amputated appendages may regenerate as appendages which belong elsewhere on the body (Needham, 1965). This strengthens the idea that the differentiated states are fairly stable and that something quite drastic has to be done to change a cell from one state to another. By contrast there are several examples from experimental embryology in which the specification of tissue can be altered merely by grafting from one position in the embryo to another, or by some alteration to an *in vitro* culture medium, some of which were mentioned in Part II.

These characteristics of differentiated cells, and the suspected but less well founded properties of embryonic cells are summarised in Table 8.1.

Dynamical systems theory

If we are thinking at the biochemical level then in order to define the state of a cell properly we need to have a complete list of all the chemical substances present with their concentrations, and if there is regional inhomogeneity we need to know the concentrations in each compartment and also the diffusion constants for the movement of each substance between each pair of compartments. Evidently a description at this level does not exist for any cell at the present time and only for bacterial cells is it even a realistic possibility in the next decade or two. So in practice we are bound to settle for something less.

The simplified definition which is at present most popular is one which ignores all substances in the cell except the genes and also ignores all compartmentation. The state of the cell is then defined as a list of those genes which are active, and if there are degrees of activity, how active they are. The idea is that the genes are ultimately the source of all biosynthesis and so the other characteristics of the cell are automatic consequences of the state of the genome, no extra information being given by stating them explicitly. This proposition is normally accepted as dogma not requiring any experimental support. In fact the best evidence both for and against it comes from some quite old experiments by Boveri on haploid hybrid merogones in sea urchins (reviewed by Morgan, 1927; Davidson, 1976). In sea urchins it is often possible to fertilise cytoplasmic fragments of the egg of one species by the sperm of another. The resulting embryos tend to resemble the maternal species up to about the prism stage and then gradually change so that they come more and more to resemble the paternal species. So in these animals at least, the assumption that cell state equals gene activity is probably valid for the later stages of development but seems less likely to hold for the early stages. As we have seen, the reason for this is that in early embryos there is often some cytoplasmic localisation of substances which determine some elements of the body plan and there is also often little genomic activity, most protein synthesis being directed by maternal mRNA.

Even if the state of a cell can be defined by a list of gene activities it is probable that the majority of gene products are not yet identified and so for practical purposes a still further simplification is necessary: this is to make the assumption that the substances in the cell belong to a number of metabolic subsystems which interact only weakly with one another. If the decisions involved in regional specification could be located in one subsystem then all the other substances could be ignored, at least to a first approximation.

The practical biochemist must assume the truth of this assumption since if regional specification cannot be ascribed to a reasonably small subset of substances then there will probably never be a biochemical solution of the problems. However, the theoretician need not worry. From his point of view it does not matter whether the state is defined in terms of a few substances, just the genes or the whole lot; in any of these cases the state will consist of a list of values $c_1, c_2, \ldots, c_n$ which represent the chemical concentrations of each of the n components which are to be included. Because the n substances undergo many interactions with one another the state of the cell is liable to spontaneous change and the direction of this change is governed by the particular laws of motion, or dynamics, appropriate to the system in question.

Continuous dynamics

According to the law of mass action the rate of increase of each chemical concentration will depend on the concentrations of the precursors. For example if $A + 2B \rightarrow C$ then $dc/dt = k_1ab^2$, where k_1 is a rate constant and a, b, c are concentrations of A, B, C respectively, If C is removed by reaction with E in the process $C + E \rightarrow F$, then the rate law for C becomes $dc/dt = k_1ab^2 - k_2ce$. In addition the rate of formation of C might be regulated by other substances which do not participate in the reaction; for example the rate might be reduced in inverse proportion to the concentration of a negative regulator G. So the full scheme would be:

$$A + 2B \longrightarrow C \longrightarrow F$$

G   E

and the rate law for C might in the simplest case be:

$$\frac{dc}{dt} = \frac{k_1ab^2}{g} - k_2ce \tag{1}$$

where g is the concentration of G. In certain applications we might need also to consider diffusion of C in and out of the compartment in question and this would necessitate adding a diffusion term to the equation:

$$\frac{\partial c}{\partial t} = \frac{k_1ab^2}{g} - k_2ce + D\nabla^2c \tag{2}$$

D is the diffusion constant of C and ∇^2 is the Laplacian operator indicating the second derivative of C with respect to space. In fact we shall not consider diffusion further until the next chapter. As far as chemical kinetics are concerned the situation is that the concentration of each substance will change in time as a function of some or all of the other substances in the cell. The actual differential equation for each substance will depend on the details of the kinetics and so in general we can only say that the rate law will be some function of the other concentrations and that it will usually be non-linear. So if there are n substances in the cell the concentrations are governed by laws of the type:

$$\frac{dc_i}{dt} = f_i(c_1, c_2, \ldots, c_n) \tag{3}$$

($c_1, \ldots, c_n$ could of course be regarded either as concentrations or gene activities). If we could measure the concentrations then it would be possible to describe the evolution of the state of the cell by plotting all of them against time. Alternatively the state can be represented at one point in time by a single point in a *state space* whose axes are the concentrations themselves. The evolution of state is then described by movement of this point along a *trajectory* in state space.

    This can most easily be understood by considering a concrete example involving only two concentrations. With two concentrations the state space can be represented on a plane by plotting the concentration of one substance against the other. The example we shall consider is a model of the *lac* operon of *E. coli* produced by Edelstein (1972), based on data of Yagil & Yagil (1971). The *lac* operon is a system of three genes responsible for the metabolism of lactose and other β-galactosides. The two enzymes of importance for our example are the β-galactosidase, which hydrolyses the substrate, and the permease, which allows it into the cell. Their synthesis is regulated by an operator gene and a repressor. When the repressor is bound to the operator the genes are inactive. When the substrate is present it binds to the repressor and detaches it from the operator, allowing the genes to become active. The true physiological inducer is actually not lactose but its isomer allolactose.

If the concentration of both enzymes (E) is represented by E, and the concentration of the inducer/substrate (M) by M, then it can be shown that the dynamics are given by:*

$$\frac{dE}{dt} = k_1 \frac{\alpha(K_1 + M^2)}{K_1 + \alpha(K_1 + M^2)} - k_2 E$$

$$\frac{dM}{dt} = k_3 E - k_4 M \tag{4}$$

The state space is shown in Fig. 8.1 in the form of a set of trajectories drawn on the E, M plane. If a cell is suddenly brought into existence with any given values of E and M we can predict that its state will evolve along the trajectory on which it lies and will continue to evolve until it achieves one of the two possible pairs of steady-state values which in this case are $E = 1.85$, $M = 0.6$ and $E = 663$, $M = 221$. There are in fact three pairs of values at which no further change will occur (i.e. at which $dE/dt = dM/dt = 0$), but the third one, which lies between the two stable points, is a saddle point and is intrinsically unstable since trajectories lead away from it as well as towards it.

The two steady states shown in Fig. 8.1 are stable because in the neighbouring regions of state space all the trajectories converge onto them. This means that once a cell has reached such a state it will persist in it. Any small fluctuations which may occur in E and M will produce only transient changes of state because the system will spontaneously move back to the stable point. Steady states like these should not be confused with states of thermodynamic equilibrium. In a completely isolated

---

* Synthesis of the enzymes is proportional to the fraction of operators free of repressor where

$$\frac{\text{operator (free)}}{\text{operator (total)}} = \frac{\alpha(K_1 + M^2)}{K_1 + \alpha(K_1 + M^2)}$$

and $K_1 = R \cdot M^2/RM_2$ for binding of inducer (M) to repressor (R): $2M + R \rightleftharpoons RM_2$ and $\alpha = K_2(R + R \cdot M^2)$ where $K_2 = R \cdot O/OR$ for binding of repressor (R) to operator (O): $O + R \rightleftharpoons OR$. The enzymes are destroyed at a rate proportional to their own concentration $(-K_2 E)$. The intracellular concentration of inducer is controlled by the permease, hence $K_3 E$, and by destruction proportional to concentration $(-K_4 M)$.

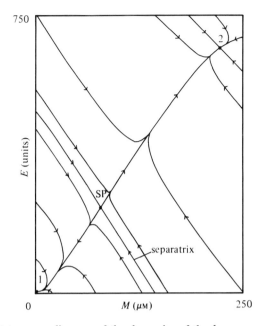

Fig. 8.1. A state space diagram of the dynamics of the *lac* operon model. Each point on the plane represents a particular pair of values of *E* and *M* and the trajectories show how *E* and *M* will change if the system is allowed to evolve spontaneously. The numbers 1 and 2 mark the stable points, and SP the saddle point.

system there would be only one steady state and this would be the composition of minimum free energy. But cells are not isolated. They are always exchanging materials with the environment and the dynamical description includes terms for this, so the steady states are kinetic rather than thermodynamic.

Since in this example there are two stable points not one, it is evident that their stability is local rather than global. Each stable point is surrounded by a region of state space called a *basin*, consisting of all the points from which trajectories converge on it. The boundary between the basins is called a *separatrix* and it is the diagonal line passing through the saddle point, marked on Fig. 8.1. If the system is in one stable state and is acted on by an external force which is sufficient to carry it across the separatrix, it will then settle into the other stable state and remain there even when the force is removed.

The model does in fact capture the essential bistable behaviour of the *lac* operon. It was shown by Novick & Wiener (1957) and Cohn & Horibata (1959) that it is possible to create populations of cells which persist in the induced or the uninduced state even when the external

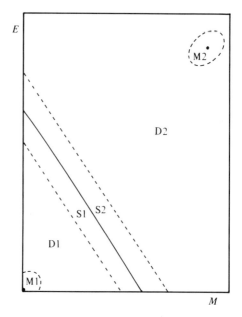

Fig. 8.2. The same state space as depicted in Fig. 8.1 could be regarded as a model for cell differentiation. The two stable points 1 and 2 now represent cell types; the zones M1, M2 represent modulations around the steady states; the zones D1, D2 represent determined states; and the zones S1, S2 represent specified states.

concentration of inducer is the same. The states are not only stable in time but also clonally heritable on cell division. It is the 'autocatalytic' property of the permease that is primarily responsible for the bistable behaviour because for a certain range of external inducer concentrations, enough inducer to derepress the operon can only enter if the permease is already present. This essential *dynamical* property of the *lac* operon has tended to be obscured in more recent years by a plethora of detailed molecular investigations and is now rarely discussed in textbooks of genetics or biochemistry.

If one is seeking a biochemical understanding of development then the language of dynamical systems theory seems a natural one to use. The example we have considered relates to a particular system of gene regulation in bacteria, but we could also regard it in a general sense as being a model for differentiation in cells of an animal embryo, and the state space is redrawn for this purpose in Fig. 8.2. The steady states can be identified with the stable *differentiated* cell types, since they are chemical compositions which are stable over time. A region of state space around each steady state is the range of *modulation* of cellular behaviour during adult life. Only very rarely will changes occur which are large enough to change one cell type into another (metaplasia), since one or more of the system variables must be altered enough to cross the separatrix. A larger region of state space corresponds to *determination*, since this is the group of states which inexorably run down to a given steady state for any environment present in the embryo. The region of state space near the separatrix corresponds to the zone of *specification*. In isolation, cells with these states will run down to one of the steady states but if exposed to inductive signals then E or M may be changed enough to cross the separatrix and run down to the other steady state. The *potency* of a cell population corresponds to all of the steady states present in the state space, since all can be reached under some condition or another.

To see how this description applies to an actual example of cellular decision-making, consider the use of this diagram to provide a representation of neural induction in a vertebrate embryo. In Fig. 8.3 is shown the situation when the ectoderm has been formed. The states of the ectoderm cells are initially all similar and situated in the bottom left of the state space. If isolated from the rest of the embryo at this stage they will all eventually become epidermis. Suppose that the neural inductor is in fact the substance M. At gastrulation M is increased in those cells receiving the signal, which are those overlying the archenteron roof. This increase changes their state in such a way that they now lie across the separatrix in the zone of neural determination and so they will eventually become neural plate. No zone of neural specification is shown in this diagram since as far as we know the change is not reversible by any environment present in the embryo. In this example an increase in E would also serve to drive the cells into the neural basin despite the fact that M and not E is the natural inductor. This example should warn us that inductive signals need not necessarily be very specific since the degree of specificity will depend on the particular dynamics in question.

Our example has dealt with a cell in which the 'attractors' are stable points on which the trajectories terminate. Another type of attractor which can be found in a two-dimensional system is the *limit cycle*, which is a closed trajectory around which the system will travel in perpetuity (Fig. 8.4). This is a state of sustained oscillation in the concentrations of both substances, and such states are not unknown in terminally differentiated animal cells – for example the intrinsic rhythmic contraction of cardiac muscle, or the continual cell division among proliferative subpopulations in the renewal tissues. The theory of limit cycles is extensively discussed by Andronov, Vitt & Khaikain (1966) and applications of the theory to biology by Pavlidis (1973) and Winfree (1980). There is no real evidence that biochemical oscillators play any role in development, but we shall consider a possible role in Chapter 10 in connection with the formation of repeated structures.

Of course real cells are likely to have thousands of interacting substances rather than two. Our example used two simply in order to be

(a)

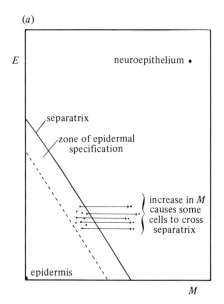

(b)

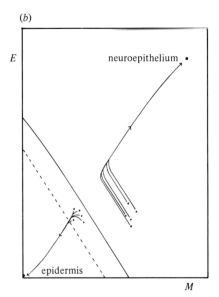

Fig. 8.3. In these two diagrams the same dynamical system serves as a model for neural induction. In (a) the cells arise in the zone of epidermal specification and are represented as a small cloud of system points since they are not exactly identical to one another. Those exposed to the inductive signal, a rise in M, move across the separatrix into the zone of neural determination. In (b) the subsequent trajectories of the two cell populations, which are the same as in Fig. 8.1, take them respectively to epidermis and to neuroepithelium.

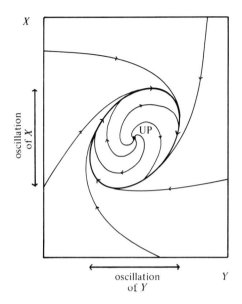

X

oscillation
of X

oscillation
of Y

UP

Y

Fig. 8.4. A limit cycle is a closed loop in state space corresponding to sustained
oscillation of the system variables. This one is a stable cycle since trajectories
converge onto it from all surrounding points. In its interior lies an unstable steady
state UP.

able to represent the state space in two dimensions on a plane. If there are
n substances in the cell then the dynamics will be representable as a system
of n equations:

$$\frac{dc_1}{dt} = f_1(c_1 \ldots c_n)$$

$$\vdots$$

$$\frac{dc_n}{dt} = f_n(c_1 \ldots c_n) \tag{5}$$

The state space has n dimensions, one axis for each chemical concentra-
tion, and the state $[c_1 \ldots c_n]$ and the rate of change vector $[\dot{c}_1 \ldots \dot{c}_n]$
each have n components. Although it is not possible to visualise a space
with more than three dimensions the principles are not very different
from those discussed above. The steady states are still points inscribed
in the n-dimensional space, and limit cycles are still closed loops although
more than two variables may oscillate as the system travels round the
loop. The space is still partitioned into basins each having one attractor,
although the separatrices are now surfaces of dimension $n - 1$. There are
in fact also some other types of attractor, more complex than the stable

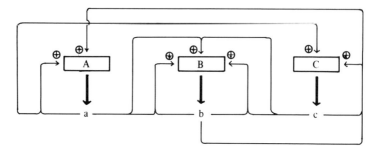

Fig. 8.5. A discrete model for gene regulation as expressed in equations (6). A, B, C are genes and a, b, c their products. The plus signs indicate sufficient conditions for activation.

point and limit cycle, which may be found in systems of more than two dimensions.

Discrete systems

The formalism of dynamical systems theory seems appropriate for the analysis of biochemical networks but there is a serious obstacle to theoretical progress. The equations of chemical kinetics, including enzyme kinetics and gene regulation, are nearly always non-linear, and mathematical techniques for solving non-linear equations are, and will probably remain, rudimentary. For this reason certain theorists, such as Thomas (1973) and Kauffman (1971, 1975) conclude that a more appropriate type of mathematics for describing the cell is that of Boolean algebra. Here all the variables are binary, taking the values 0 or 1, and the functions relating them are logical equations. This approach is inherently non-linear and much simpler since it is possible to solve problems involving many variables by analytical methods. It also lends itself to the reduction of the cell state to a list of gene activities because each gene can be identified with a single binary variable taking the value 0 ('off') or 1 ('on'), and can be expressed as a function of the activities of the other genes.

Let us consider another example of a bistable circuit shown in Fig. 8.5. This shows three genes A, B and C, whose activity is regulated by the respective gene products a, b, c. A is 'on' if a is present or if both b and c are present. C is regulated in the same way. B is 'on' if any two of the gene products are present. In Boolean algebra:

$$
\begin{aligned}
A &= a \; or \; (b \; and \; c) \\
B &= (a \; and \; c) \; or \; (a \; and \; b) \; or \; (b \; and \; c) \qquad (6) \\
C &= a \; or \; (b \; and \; c)
\end{aligned}
$$

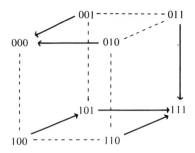

Fig. 8.6. The state diagram of the system of Fig. 8.5. There are two basins draining to the stable points 000 and 111 respectively.

Note that the state of the genes is represented by one set of variables (A, B, C) and the presence of the gene products by another set (a, b, c). It is this that introduces a time dimension since the products appear after their genes have been activated. In fact the states of activity are analogous to the time derivatives in equations (1) to (5). The state space consists of eight discrete points representing all the possible configurations of gene activity. For example, '000' means that all three genes are 'off'; 010 means that B is 'on' but A and C are 'off', and so on. The states are conveniently shown as a cube (Fig. 8.6), since neighbouring states can then be placed adjacent to one another in the diagram. Each state will spontaneously evolve according to the dynamics of equations (6) and this leads to the transitions shown. There are two stable points, 000 and 111, in which the system will persist indefinitely. The rest of the space is partitioned into two basins in which the trajectories run down to the two stable points. We could, if we wished, again draw a separatrix between the two basins. Again we can regard the stable points as representing the stable cell types and the remainder of each basin as representing committed states: specified if they can be altered by environmental stimuli present within the embryo and determined if they cannot. If we choose to regard this as another model of neural induction then the ectoderm could arise in the state 010. On its own this will develop to 000 (epidermis). Suppose that the neural inductor is the gene product a. If enough exogenous substance is applied to some of the cells to cause A to change from 0 to 1 then the overall state will become 110 and this will then evolve spontaneously to 111 (neural plate). This effect could also be brought about by treatment with the substance c which is not the natural inductor.

The discrete formalism is not exactly equivalent to the continuous. The definitions of stability are not quite the same, the continuous system can have more complex types of attractor than the point and the cycle, and since time is discrete in the logical equations it is necessary to augment the system with rules about the order in which variables change when more than one change is necessary in one time step. However, there are many similarities between discrete and continuous treatments and one representation may be chosen over the other depending on the ease of computation in a particular case.

The point of this section has been to show that if we are serious about attempting to understand the hierarchy of developmental decisions in biochemical terms, then we do not just need to identify the relevant genes and gene products but also to understand their dynamical behaviour. In the past this has proved to be necessary for understanding such things as the mechanism of nerve conduction and aggregation in slime moulds. It seems probable that in the future it will be through the mathematics of dynamical systems theory that embryological and biochemical results will be brought together in a productive way.

Complexity of cell states

Until recently it has not been possible even to guess at the real complexity of the cell at the biochemical or genetic level. But in the last ten years or so measurements have begun to be made which give us at least an order of magnitude estimate of the problem. There are three sorts of measurement: counting genes, counting mRNAs and counting proteins. In principle the last two methods give minimum estimates of complexity since they only count genes which are 'on' in the cell type or stage examined. From a regulatory point of view what is 'off' may be just as important as what is 'on'.

The best estimate for the gene number in an animal is 5000 for *Drosophila melanogaster*. This was obtained by isolating a large number of lethal mutations from a small chromosome region and comparing the number of complementation groups with the number of cytologically visible bands (Judd & Young, 1973). The numbers are very similar and since there are about 5000 bands altogether we get the estimate of 5000 genes. This is far fewer than the number of genes which could be formed if all the genomic DNA was active, but it is now thought that probably a majority of the DNA is not active since it consists of blocks of repeated sequences which often differ in a striking fashion between related species (Lewin, 1980). The number of active genes in other animals is not really known since estimates must rely on extrapolation from very small regions which have been mapped by genetic or physical methods. Bodmer (1981) has argued that in higher animals the *cluster* of related genes is the basic unit and for the human the total gene number may be in the vicinity of 3000 to 10 000, this being the number obtained by dividing the non-repeated DNA by the average cluster size. In view of the estimates derived from counting mRNA sequences this may be somewhat on the low side.

The most detailed measurements of mRNA complexity have been made by Galau *et al.* (1974, 1976) using the sea urchin embryo. They find that about 1.7×10^7 nucleotides-worth of different sequences exist in mRNA isolated from the polysomes of gastrulae. If the average protein contains 500 amino acids then this must represent something of the order of 10 000 structural genes. Oocytes and blastulae contained most of these plus about as many again of different ones, while adult tissues contained 2000–4000 genes-worth of mRNA of which about half were found among the gastrula sequences and half were not. Tissues and cell lines of mouse, chick and *Xenopus* have yielded mRNA complexities in the range of 10 000–30 000 genes-worth (Davidson & Britten, 1979). There is some uncertainty about these estimates since most of the sequences are only present in one to fifteen copies per cell and so small errors of measurement make large differences to the results.

The best way of counting proteins is to count the spots on well-resolved two-dimensional gels. This yields a value of up to about 1000 for any given cell type, but is obviously a minimum estimate since rare species will not be visualised in the presence of abundant ones. One thousand is about the number of moderately prevalent mRNA species, that is species present at 15–300 copies per cell, and so we are probably looking at the products of these messages when we look at two-dimensional gels.

We are left feeling that the minimum number of active genes in a particular type of cell is around 1000 and the maximum number perhaps 100 000. Which of these one chooses to believe in has a certain relationship to one's attitude to the process of explanation. A thousand proteins could all be named, isolated and described whereas 100 000 could not. Furthermore, it is important from the developmental biologist's point of view to know whether the dynamical behaviour that he or she is interested in is manifested by a small subset of these substances in relative isolation from the remainder of the cell, or only by the whole thing. If it is a small subset then all the relevant components could be isolated and the dynamical behaviour could be worked out by experiments *in vitro*. However, if it is the whole thing, and particularly if the whole thing consists of 100 000 components, then there will never be any reductionist explanation of developmental phenomena. The best that could be hoped for is the deduction of general rules by studying the behaviour of model networks using the techniques of dynamical systems theory. At the time of writing the question of numbers is still open and so the developmental biologist is free to believe what he or she likes.

Symmetry-breaking processes

One of the questions addressed at some length by theoretical biologists is how a uniform initial state can spontaneously change into one with regional differentiation. Where the environment of the tissue is asymmetrical there is no problem, but what of an entire embryo suspended in a fluid of uniform composition? Although few, if any, animal eggs are truly uniform when they are laid, there are some cases in early embryogenesis where polarisation seems to arise *de novo* – for example in the position of the blastocoelic cavity in the mouse embryo, or the formation of the dorsoventral axis in the amphibian embryo. This sort of symmetry-breaking process seems to violate the conception presented above of the cell as a deterministic dynamical system, insofar as there is no obvious cause for the orientation of structures.

The answer really lies in distinguishing between what Aristotle called the 'formal cause' and the 'efficient cause' of an event. In answer to the question 'Why does the light come on?' we might answer that given the mains voltage and the resistance of the lamp filament, the energy dissipation is such as to make the filament incandescent. We might alternatively answer that it was because someone flipped the switch. The first is the formal cause, roughly the laws of motion of the system; and the second is the efficient cause, or the agent which unleashes a particular process.

In the present context we can distinguish a formal and an efficient cause for symmetry-breaking in systems made up of chemical reactions. The formal cause lies in the dynamics of the system, which must be of a type involving bifurcation of the steady-state behaviour. This means that the system is governed by some parameter such that above a certain threshold one stable state bifurcates into two. The efficient cause lies in the molecular nature of matter which means that at a given point in time a well-mixed solution will have very small regional differences in composition arising solely from the random distribution of molecules. The symmetry-breaking occurs when one of these microscopic perturbations is amplified by the dynamics into a macroscopic inhomogeneity. An example is provided by the celebrated 'Brusselator', which is a model reaction system studied in detail by the physical chemists of Brussels (Nicolis & Prigogine, 1977). The reactions are as follows:

$$A \rightleftharpoons X$$
$$B + X \rightleftharpoons Y + D$$
$$2X + Y \rightleftharpoons 3X$$
$$X \rightleftharpoons E$$

The kinetic equations are:

$$\frac{\partial X}{\partial t} = A - (B + 1)X + X^2Y + D_x\nabla^2X$$

$$\frac{\partial Y}{\partial t} = BX - X^2Y + D_y\nabla^2Y$$

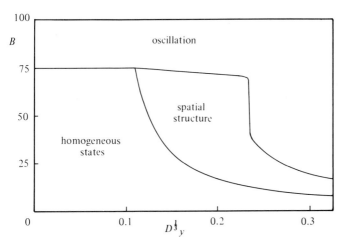

Fig. 8.7. Behaviours of the 'Brusselator' as a function of the parameters B and D_y. In this particular example $D_x = 1.05 \times 10^{-3}$ and A is assumed diffusible with boundary concentration 14 and diffusion constant 0.195.

These describe the rates of change of concentration of the intermediates X and Y (X and Y respectively) in terms of the law of mass action and of diffusion. The Laplacian operator ∇^2 is a second derivative in space and so the terms $D_x\nabla^2 X$ and $D_y\nabla^2 Y$ represent the parts of the rates of change of concentration which are attributable to diffusion. B, D_x and D_y are parameters for the system. This means that each time the equations are solved B, D_x and D_y are treated as constants, but if their values are changed the solutions (in this case the steady-state behaviour of X and Y) may be different.

In Fig. 8.7 are shown domains of B and D_y in which the system exhibits particular behaviours, B playing the role of the bifurcation parameter. If we start with $B = 25$, $D_y^{1/3} = 0.1$, then there is a single stable steady state. If D_y is allowed to rise then there will soon come a critical value above which this steady state becomes unstable and a spatially non-uniform state becomes stable. This can roughly be thought of as a situation similar to that shown in Fig. 8.1 in which there are two stable steady states, one with high concentrations of X and Y and the other with low X and Y, except that here the two states coexist in the same reaction vessel and X and Y diffuse from regions of high concentration to regions of low concentration. The structure arises because it is selected for amplification from myriads of minute transient asymmetries generated by molecular motion. If there are alternative outcomes which differ only in polarity, for example whether a high point is to the left or to the right, then the one selected will depend on any slight asymmetry which is present in the initial conditions, even if it is imperceptible to normal measurements and not directly involved with the mechanism. If we start from the point $B = 25$, $D_y^{1/3} = 0.1$ and increase B there will again come a critical value beyond which the steady state breaks down, this time to be replaced by a sustained oscillation of X and Y, in other words a limit cycle. In this case high X/low Y leads to low X/high Y because X needs Y to maintain its high level, otherwise it will decay to E and Y. Low X/high Y leads back to high X/low Y because Y can only decay to X.

The Brussels school have shown that replacement of homogeneous by inhomogeneous steady states is something which can occur only far from thermodynamic equilibrium. This means that there must be a considerable flux of material through the reaction network – exactly as found in cells and whole organisms. The inhomogeneous patterns which emerge are called 'spatial dissipative structures', to emphasise that as open systems they are continuously dissipating energy.

We shall return to dissipative structures in the next two chapters when we consider the regulation of gradients and the formation of repeating patterns. For the moment the important point is that there is no fundamental problem about symmetry-breaking. Certain types of kinetic network can be guaranteed spontaneously to generate structure and the orientation of the structures will be determined by tiny and imperceptible pre-existing asymmetries. This conclusion is probably one of the most important results of theoretical biology to date.

9

Gradient models for embryonic induction and regulation

It should be clear from the way in which embryological phenomena have been described in this book that we do not wish simply to enumerate the cell states present in the adult, nor simply to describe the trajectories which cell populations follow in their state space, but above all to account for the assignment of particular states to different regions of the early embryo. It is the understanding of this process, called 'regional specification' here, that would come nearest to answering the complex and global question: 'How does the egg become the organism?'

The models which will be discussed in this and the following chapter deal with this problem. In the present chapter we shall consider gradients and in the final chapter non-gradients. A 'gradient' to the embryologist may best be defined as a smooth and monotonic variation of cell state with position. 'Monotonic' means no maxima or minima of the variables in question, which will usually, but not always, be chemical concentrations. To earlier theorists such as Child (1941) the gradients were of 'metabolism', while to contemporary ones such as Crick (1970) they are diffusion gradients of single substances, but in either case their presumed role is as vehicles of inductive signals from organiser to responding regions. Those who have a nodding acquaintance with embryology often seem to regard gradients as some form of mystical pseudoexplanation and a distraction from the serious business of biochemical analysis. In the present chapter this prejudice will be tackled by attempting to show to what extent the various models are really capable of explaining the features of early embryonic behaviour summarised in Chapter 7, and to what extent they are deficient.

Diffusion of a morphogen

For ease of computation, models involving diffusion are usually one-dimensional, referring either to diffusion along a line of cells or across a sheet of cells in which all the sources and sinks are arranged as transverse rows.

If two solutions containing different concentrations of a substance are placed in contact then the material will start to diffuse from the region of high concentration to the region of low concentration and the transport across a unit area is given by

$$\text{Flux, } J = -D \frac{dc}{dx} \text{ mol} \cdot \text{s}^{-1} \tag{1}$$

where dc/dx is the local concentration slope and the minus sign indicates that the flux is down the gradient; and D is the diffusion constant and is measured in units of $cm^2 \cdot s^{-1}$. The larger the value of D, the faster the substance spreads; the D depends both on the characteristics of the molecule in question and on the viscosity of the medium. If cell membranes have to be crossed the diffusion constant will be rather small for all but the most hydrophobic substances. If diffusion across membranes depends on gap junctions or carrier molecules then D can be regarded as a variable which depends on the state of the cells. Crick (1970) has estimated $D = 0.27 \times 10^{-6}$ as a reasonable value for a small molecule crossing membranes by a carrier mechanism.

If we consider the change of concentration in a cell due to diffusion we see that it must be proportional to the difference between the flux in and the flux out, in other words to the rate of change of concentration slope with position, which is the second derivative of c, hence:

$$\frac{\partial c}{\partial t} = D \frac{\partial^2 c}{\partial x^2} \tag{2}$$

i.e. (rate of change of concentration in time) = $D \times$ (rate of change of concentration slope in space)

This is the fundamental equation of diffusion. If regions containing different concentrations of a substance are placed in contact then eventually diffusion will spread the substance out so that it has a uniform concentration everywhere. However, if the substance is being produced in some places and removed in others then the distribution of material will tend towards a non-uniform steady state in which $dc/dt = 0$ at all positions and there is a continuous flow of material from the sources towards the sinks. For example, suppose that there is a source at one end ($x = 0$) of a line of cells and there is a sink at the other end ($x = l$). At the source the concentration is fixed at $c = c_0$ and at the sink it is fixed at $c = 0$. In the steady state:

$$\frac{\partial c}{\partial t} = D \frac{\partial^2 c}{\partial x^2} = 0 \tag{3}$$

therefore

$$\frac{d^2c}{dx^2} = 0$$

therefore

$$c = Ax + B$$

At $x = 0$, $c = c_0$ therefore $B = c_0$

At $x = l$, $c = 0$ therefore $A = \frac{-c_0}{l}$

So

$$c = c_0\left(1 - \frac{x}{l}\right)$$

This gives a linear steady-state gradient with the concentration falling from c_0 to 0 across the field. Munro & Crick (1971) have shown that the time required to bring c to within 1% of the steady-state value can be expressed as

$$\text{time} = \frac{Al^2}{3600D} \text{ hours} \qquad (4)$$

where A is a constant which depends on the starting conditions and has a value between about 0.1 and 0.5. As emphasised by Wolpert (1971), early embryos and embryonic fields are typically small, rarely exceeding 1 mm in linear dimensions. If $l = 0.1$ cm and $A = 0.5$ and $D = 0.27 \times 10^{-6}$ cm$^2 \cdot$s^{-1} then the time to set up a gradient is 5–6 hours. In terms of developmental time scales this seems a reasonable time to take over each decision. There are some organisms, such as *Drosophila*, or the ascidians, which develop to larvae in about 1 day, and it might be thought that diffusion provided too slow a mechanism for inductive signalling in such cases. However, it may be that the gradient need not be as close to the steady state as 1%; if it were 10% the time required would be substantially less.

The local source–dispersed sink model

Depending on how the sources and sinks are arranged, the diffusion equation has different steady-state solutions. One particularly popular model puts the source at one end with a constant concentration c_0, and the sink as the entire responding field with each cell destroying the morphogen at a rate proportional to its local concentration. There is no exit from the field so the far end acts as a barrier for the morphogen. Since there can be no flux across this barrier it follows from equation (1) that the local concentration slope, $(dc/dx)_l$, must be zero. So

$$\frac{\partial c}{\partial t} = D\,\frac{\partial^2 c}{\partial x^2} - kc$$

In the steady state

$$D\,\frac{d^2 c}{dx^2} - kc = 0 \tag{5}$$

Let $k/D = \alpha^2$ then

$$\frac{d^2 c}{dx^2} - \alpha^2 c = 0$$

This is a well-known type of differential equation to which the solution is

$$c = Ae^{\alpha x} + Be^{-\alpha x}$$

The boundary conditions are

$$c = c_0 \text{ at } x = 0$$

$$\frac{dc}{dx} = 0 \text{ at } x = l$$

and when these are substituted we arrive at

$$\frac{c}{c_0} = e^{-\alpha x} + \beta \sinh \alpha x \tag{6}$$

where

$$\beta = \frac{e^{-\alpha l}}{\cosh \alpha l}$$

(sinh and cosh are the so-called hyperbolic functions whose values can be looked up in tables)

The solution is sketched in Fig. 9.1 for three values of α, showing that it is a monotonic gradient which is steeper the larger the value of α.

If this gradient were an inductive signal we could assume that it would turn on biochemical switches in the responding tissue at certain threshold concentrations. We shall not consider the mechanism of thresholds until the next chapter, but since it is the positions of the thresholds that determine the anatomy of the organism it is clear that disturbances of the steady-state gradient would produce, one to one, disturbances in the anatomy. Let us now consider the effects of certain embryological manipulations of a type which have been used on real embryos and discussed in Part II.

If the source region is left fixed and other regions of tissue are rearranged, then the gradient will eventually relax back to its original configuration (Fig. 9.2). Similarly if a small region is removed which is

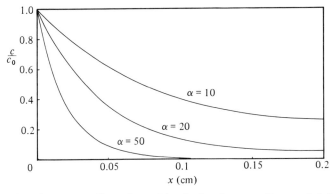

Fig. 9.1. Steady-state gradients formed by the local source–dispersed sink model (equation 6). $\alpha = \sqrt{(k/D)}$ and determines the steepness of the gradient. In this case the length of the field, l, is 0.2 cm.

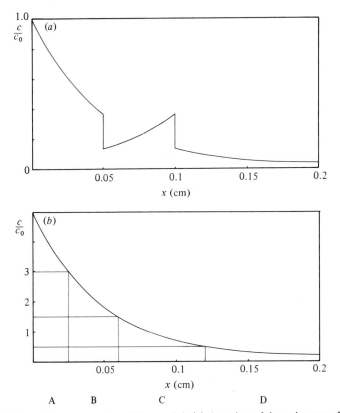

Fig. 9.2. Regulative properties of the model. (*a*) A region of tissue is rotated (0.05 $< x < 0.1$) but the gradient will relax to its steady state and if it does so before the threshold responses of the cells shown in (*b*), then the order and proportions of zones A, B, C, D will be normal.

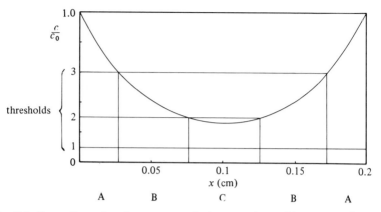

Fig. 9.3. Formation of a mirror-symmetrical pattern by grafting a second source region at $x = 0.2$. The new steady-state gradient is U-shaped and the same threshold responses shown in Fig. 9.2 will produce the anatomy ABCBA.

fated to form a certain structure, then the structure will form nonetheless from the tissue on either side. As long as the threshold responses do not take place until the steady state has become re-established this type of simple gradient model provides a satisfactory explanation of defect regulation.

It also explains the special properties of the organiser regions by identifying them with sources. If a source region is removed then the steady-state concentration will fall to zero and no structures will be formed. This behaviour is shown by all the putative organisers of Table 7.3 except for Hensen's node, the amphibian organiser and the sea urchin micromeres. If the source is moved from one end of the field to the other then the structures will be formed in the correct sequence but with an overall reversal of polarity. If a second source is grafted to the far end of the field then a duplication is formed (Fig. 9.3). This consists of two sets of similar structures which are arranged with opposite polarity around a central plane of mirror symmetry. This type of mirror duplication has been encountered several times: as an outcome of organiser grafts in amphibians, following equal cleavage in the annelid *Chaetopterus*, induced by ultraviolet irradiation in the midge *Smittia*, and as an outcome of the maternal-effect mutant *bicaudal* in *Drosophila*. The gradient model explains the overall symmetry, the fact that normal neighbour relations are maintained between structures and the fact that some structures are missed out because the central minimum of the U-shaped gradient exceeds some of the lower thresholds. It does not, however, explain how the second source arises.

If we assume that the time at which the switches are activated is later in regulative than in mosaic embryos then the same type of model can be

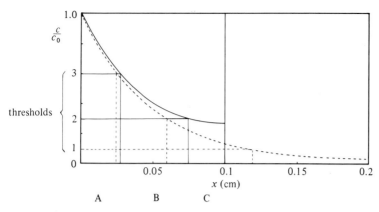

Fig. 9.4. Failure of the simple gradient model to regulate its proportions in a domain of reduced size. The field has been reduced in size from 0.2 to 0.1 cm and the steady-state gradient must regulate upwards since there can be no flux across the new boundary. The same thresholds are operative as in the previous figures and produce only zones A, B and C with boundaries shifted to the right.

used to cover both. The cytoplasmic localisations collected in Table 7.2 can be regarded either as arising from the early activation of switches or, in the cases where the regions concerned are themselves organisers, as incipient sources. Since the essential events of specification are regarded as cytoplasmic and metabolic, it is not surprising that nuclei of early embryo cells seem to remain totipotent for some time after the cells themselves are committed. Stabilisation of differential gene expression will follow cellular determination.

By and large the simple gradient model is able to account for many of the facts of early development, but there are some phenomena which seem at first sight to be incompatible with it. There are three of these: adaptation of proportions to overall size, symmetry-breaking and regulation of organisers.

The problem concerning proportion regulation is illustrated in Fig. 9.4. If the size is reduced, for example by insertion of a barrier in the midst of the field, then the morphogen accumulates on one side of the barrier and disappears on the other. When the threshold responses are drawn in, it is clear that nothing is formed on the far side and the structures formed on the near side are shifted towards the barrier relative to the positions of their presumptive areas on the normal fate map.

In fact this is similar to what is found in the so-called gap phenomenon in the eggs of insects such as *Bruchidius* (see Chapter 4). Taken together with the production of mirror duplications in *Drosophila* and *Smittia* this provides good evidence for the existence of some sort of gradient in the early insect embryo. However, the phenomenon is not a general one, since in most types of embryo reduction of size at an early enough stage leads to the formation of a small but normally proportioned larva. Also, when a symmetrical duplication is formed, the simple gradient model predicts that half or less than half of the fate map should be expressed in each half of the duplication (see Fig. 9.3). However, in the case of organiser grafts in *Xenopus* it seems that *more* than half of the fate map is expressed on each side (Cooke, 1981). This suggests that the central minimum of the U-shaped gradient is a *lower* concentration than would arise at this position in normal development, and this cannot be explained by a simple diffusion model.

In several cases it is also possible to *augment* the size of embryos and obtain large normally proportioned body plans (Table 7.1), and the model does not predict this either. It is true that the model which has a source at one end of the field and a sink at the other would permit size regulation, but the sink would then also behave as an organiser and there is no evidence for an organiser at both ends of the field in any of the examples we have considered.

Secondly, there is the matter of symmetry-breaking. We have seen that there is some reason for thinking that some regional differences are set up in response to stimuli which are very minor environmental perturbations. Examples of this are the orientation of the dorsoventral axis in the amphibian egg, or of the craniocaudal axis in the isolated anterior half of an avian blastoderm. The simple gradient models require some pre-specification of the sources and sinks, which seems unlikely to exist in such cases.

Thirdly, there is the regulation of organisers. Of the organisers listed in Chapter 7 three can be removed without severe effects on the subsequent pattern, namely the early blastopore lip in the amphibian, Hensen's node in the chick and the micromeres of the sea urchin morula. If the others are removed the pattern of structures formed by their fields of influence is severely defective.

These objections are serious enough to have caused some workers to produce more elaborate gradient models such as those described below. While they may be right it is worth considering whether the case against simple gradient models really is overwhelming or whether there are ways around it. The argument about size regulation is based on the *steady-state* properties of the diffusion gradient. However, as mentioned above in the context of the timing of events, we do not know whether diffusion gradients in embryos ever reach a steady state. If they do not then there is no problem about size reduction since there will be a transient state in which the gradient spans its normal range of concentrations, and if the threshold responses occur at this stage all structures will be represented in the anatomy. This argument is no use in the case of *augmented* size, but here it is possible to argue that the double-size embryo is in any case an illusion, and that all that has really happened is that the extra-embryonic portion has increased in size. All embryos have an extra-embryonic portion even if it is quite small, such as that part of the amphibian endoderm destined to end up within the gut lumen; and we know from the careful studies on the early mouse embryo that the proportions are not necessarily maintained with regard to embryonic and extra-embryonic regions. It will take quite a lot of careful research on the proportions of embryos following size changes before we can really be sure that the simple gradient model cannot explain the facts.

Symmetry-breaking is an awkward problem because we can never be sure that it is really happening. All animal eggs are polarised in some way when they are laid and though in a few cases they seem to be symmetrical with respect to the establishment of some axes it is not possible to prove the absence of a significant internal asymmetry which normally sets in train the formation of a source region.

As for the matter of the regulation of organisers, in no case do we know the exact spatial extent of the organiser and so it is always possible to argue that part of it remains. In the cases mentioned, defective embryos do arise if a large enough piece of tissue is removed from the organiser region.

To conclude: in the case of the early insect embryo the local source–dispersed sink model gives a good qualitative explanation of the results. In other cases it fits some but not all the data and at present the data are not good enough for us to reject it decisively.

Gierer and Meinhardt's double-gradient model

The most exhaustive study of double gradients has been carried out by Gierer & Meinhardt (1972) and Meinhardt & Gierer (1974). Because of the difficulties of analytical non-linear mathematics alluded to in the previous chapter they have used the method of computer simulation to investigate a variety of models and apply them to a number of situations in developmental biology. From an explanatory point of view their models are superior to the simple gradient models in three respects: they have symmetry-breaking properties and so can establish a gradient across an initially homogenous field, they can reconstitute the source region if it is removed, and they have some ability to regulate in the face of changes in size of the field.

Their first model depends on *lateral inhibition* and involves two morphogenetically active substances called the activator and the inhibitor. The activator (A) stimulates its own production (autocatalysis) and also that of the inhibitor (I), while the inhibitor represses the formation of the activator. Both substances are removed at a rate proportional to their concentrations (A and I respectively).

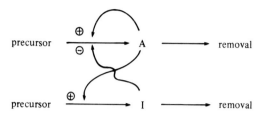

In addition both substances are diffusible, and the inhibitor diffuses faster than the activator. The kinetic equations for the system in one dimension are:

$$\frac{\partial A}{\partial t} = k_1 \frac{A^2}{I} - k_2 A + D_A \frac{\partial^2 A}{\partial x^2} \tag{7a}$$

$$\frac{\partial I}{\partial t} = k_3 A^2 - k_4 I + D_I \frac{\partial^2 I}{\partial x^2} \tag{7b}$$

As in previous examples the final terms represent diffusion according to equation (1), and the first and second terms represent production by chemical reactions. In some versions there is a low basal level of activator production as well. This system is similar to the 'Brusselator' described in the previous chapter in that in a certain parameter range the homogeneous state is unstable and spontaneously breaks down into an inhomogeneous state in which there are regions of net production of activator and inhibitor (sources) and regions of net removal (sinks). It is easy to understand how this happens. Suppose we start from the unstable homogeneous state (in which $A = k_1 k_4 / k_2 k_3$ and $I = k_1^2 k_4 / k_2^2 k_3$). If a slight perturbation should increase the local concentration of A, then because both substances are made at a rate proportional to A^2, there will be a rapid net local increase of both A and I. This will establish a concentration gradient and both substances will diffuse towards the surrounding regions in which their concentrations are lower. But I diffuses faster than A, so A will then predominate over I at the original disturbance centre and I over A in the surroundings. Because of the autocatalysis the inhomogeneity will grow until a steady state is reached in which local production of A in the source region (as it has now become) is balanced by its removal by diffusion. The steady state will show a sharp peak of activator concentration and a rather broader one of inhibitor concentration (Fig. 9.5). The number of activator peaks which may arise depends on the total size of the field relative to the activator diffusion constant, but in the present chapter we shall consider only situations where the normal number of peaks is 1 and the normal pre-pattern consists of a monotonic gradient of A and I.

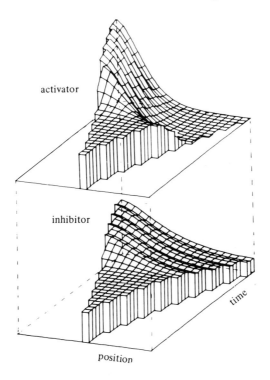

activator

inhibitor

time

position

Fig. 9.5. The lateral inhibition model of Gierer and Meinhardt. This figure shows the concentrations of activator and inhibitor at different times in a growing domain. Once a certain critical size has been reached the homogeneous situation becomes unstable and a peak forms at one end of the field. The peak of the inhibitor is broader than that of the activator because its diffusion constant is greater. (Figure kindly provided by Dr H. Meinhardt.)

In the field of early embryogenesis the lateral inhibition model has been applied in the greatest detail to insect development (Meinhardt, 1977). It is assumed that the inhibitor is the actual morphogen and that in normal development the source region forms at the posterior pole (the activation centre). We have already seen that several features of early insect development can be explained by the simple gradient model, namely the gap phenomenon and the inversions of polarity produced by displacement of the posterior cytoplasm in *Euscelis*. The Gierer–Meinhardt model explains both of these along similar lines but can also explain the production of double abdomens in *Smittia* by ultraviolet irradiation. The assumption here is that the irradiation destroys the inhibitor. So if the anterior region is irradiated, a new source region is established because the local inhibitor concentration is depressed and so the activator increases autocatalytically. As the new activated region grows it becomes

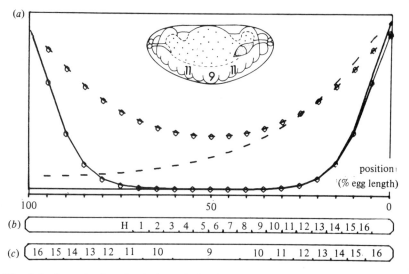

Fig. 9.6. Quantitative simulation of the formation of the double abdomen in *Smittia* embryos. (*a*) The normal and U-shaped gradients of activator (continuous lines) and inhibitor (dashed lines) along the egg. (*b*) The normal positions of the prospective regions for the segments on the fate map. H, head. (*c*) The positions of segments expected in the duplicated pattern. Inset in (*a*) is a diagram of a double embryo with segment 9 at the mirror plane. (Figure kindly provided by Dr H. Meinhardt.)

itself a source of new inhibitor and eventually a U-shaped inhibitor gradient is formed along the egg (Fig. 9.6). Conversely irradiation of the posterior region has little effect. There will in this case be a temporary local overshoot of activator followed by inhibitor but this will not affect the identities or the order of structures determined by the inhibitor gradient.

With a change of parameters the model is able to re-form a new source when the old one has been removed, and the new source will always form at the end of the fragment which was nearest to the old source. This is because it is at this point that there is the greatest predominance of activation over inhibition. Organisers which appear to some extent to re-form after removal are the dorsal marginal zone in the amphibian embryo, Hensen's node in the chick embryo, and the micromeres of the sea urchin.

In addition the model can accommodate a certain degree of change in the overall size of the embryo, both in terms of retaining the capacity to form a monotonic gradient and to some extent preserving the proportions of the embryo. In Fig. 9.7 are shown concentration profiles produced on full-size, half-size and one-sixth-size fields. It is here necessary to assume

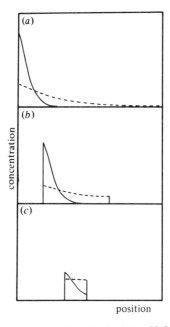

position

Fig. 9.7. Partial regulation of proportions by the lateral inhibition model of Gierer & Meinhardt. In this version there is a slight intrinsic gradient of activator production which imposes a polarity on the field. (*a*) Steady-state gradients of the activator (continuous line) and inhibitor (dashed line). (*b*) The gradients which arise if half of the field is isolated. The height of both peaks has been slightly reduced but the concentrations spanned by the activator are equivalent to most of the original field. (*c*) What happens if one third of the half-size field is isolated. A gradient is still formed but only spans a fraction of the original concentration range. (After Gierer & Meinhardt, 1972.)

that it is the activator rather than the inhibitor which acts as the morphogen. The concentration span of the activator is quite well preserved in the half-size pattern, not so well in the sixth-size. But half linear size corresponds to one eighth volume, which we have seen to be the known range of approximate proportion regulation in amphibian and starfish embryos. If the extreme concentrations determined extra-embryonic regions whose proportions were not conserved then the model could explain proportion regulation to a greater degree than is known to exist at present. Furthermore, if there is a saturation of activator production, as for example is obtained by substituting $A^2/(1 + KA^2)$ for A^2 in equation (7a), then the proportion regulation becomes even better. The problem has been studied for a different but related model by Lacalli & Harrison (1978), who showed that a monotonic gradient, once set up, was stable to changes of size up to a factor of about 10.

Other double-gradient models

The particular kinetics of equations (7) are not the only ones which show the essential properties of the double-gradient models, namely symmetry-breaking, defect regulation and proportion regulation. In fact *any* pair of chemical reactions will do so as long as they satisfy the following conditions (Gierer, 1981):

1. one component must be autocatalytic;
2. the other must inhibit the first;
3. the inhibitory effect must be strong enough to prevent an explosion;
4. the inhibitory effect must be fast compared with the autocatalysis;
5. the range of activation must be smaller than the field size;
6. the range of inhibition must be sufficiently large relative to the range of activation.

('Range' is the mean distance between production and decay of molecules.) This means that there is an infinity of possible models. It is not possible to falsify the whole class of models by experiment because if a particular one is excluded another with different parameters could always be advanced. Nor is there any guarantee that there is only one activator and one inhibitor. There may be many more substances involved in the patterning mechanism, some of which behave collectively as the activator and others which behave collectively as the inhibitor. So it is quite probable that a biochemical search for a single substance with the properties of the activator or inhibitor in equations (7) would prove unsuccessful even if the overall kinetics of the mechanism were as described. There is not even any guarantee that induction is mediated by diffusing substances. If the state of one cell can modify the behaviour of its neighbours via membrane contacts then exactly similar behaviours can be produced despite the fact that no substance passes from cell to cell (Babloyantz, 1977). This sort of inconclusiveness recurs again and again in theoretical biology and must be recognised as the major weakness of the approach.

Multicomponent gradients

A monotonic gradient of one or more morphogens must be interpreted in the form of stable states of determination by the cells of the responding tissue. How this might be done is considered in Chapter 10, but it is also possible that there is no morphogen gradient at all but rather a sequence of states of determination which can activate one another in a hierarchical manner. An example is provided by a model of Meinhardt & Gierer (1980). This model depends on lateral activation of states. Each state is defined by the activity of one gene from a set and the concentration of the

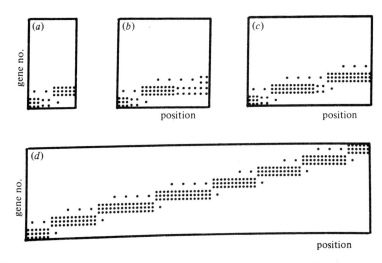

Fig. 9.8. The multicomponent gradient model described in the text and by equations (8). (*a*)–(*d*) depict different stages between which the field expands by marginal growth on the right-hand side. Each rectangle of dots indicates the concentration profile of one of the products g_i. In (*a*) and (*b*) only g_1 and g_2 are activated, in (*c*) g_3 is activated and in (*d*) g_1–g_8 are activated. (Figure kindly provided by Dr H. Meinhardt.)

ith gene product is represented by g_i. The activity of each gene is controlled by the level of its own product, by the level of a common repressor (r) and by the level of substances (s_i) formed from the products of the 'neighbouring' genes $i - 1$ and $i + 1$.[*]

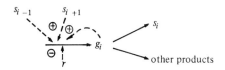

In essence this system resembles the lateral inhibition model of equations (7) in that each g_i is an activator which increases its own formation and

[*] Footnote:

$$\frac{\partial g_i}{\partial t} = \frac{c_i g_i'^2}{r} - \alpha g_i + D_{g_i} \frac{\partial^2 g_i}{\partial x^2}$$

where

$$g_i' = g_i + \delta^- s_{i-1} + \delta^+ s_{i+1}$$

$$\frac{\partial s_i}{\partial t} = \gamma(g_i - s_i) + D_s \frac{\partial^2 s_i}{\partial x^2}$$

$$\frac{\mathrm{d}r}{\mathrm{d}t} = \Sigma c_i g_i'^2 - \beta r \tag{8}$$

that of the repressor. But in addition each g_i is increased by products of the genes above and below in the sequence. In order that a *sequence* of states is set up it is necessary that there should be an asymmetry of these cross-activations: $\delta^- > \delta^+$. There is also a hierarchy of the autocatalytic constants $c_i > c_{i+1}$.

This model will establish a sequence of states in a growing domain (Fig. 9.8). Each state is characterised by the activity of only one of the genes because of the common pool of repressor: if one gene is slightly more active than the others then repression will prevail over autocatalysis for all the less active ones. The sequence of states is accompanied by a monotonic gradient of the repressor, but the repressor is not itself diffusible and cannot be considered a 'morphogen' in the sense defined earlier. This particular model is able to re-fill gaps in the pattern and variants of it establish patterns on non-growing fields and regulate proportions over a threefold range of linear dimensions.

These models still correspond to the definition given at the beginning of this chapter – 'a gradient is a smooth and monotonic variation of cell state with position' – if the state is expressed in terms of the gene index i. However, the activator concentrations (g_i) are no longer distributed monotonically. Also, although the absence of sharp discontinuities is maintained by diffusion there is no single substance that diffuses from one end of the field to the other. It is important to emphasise that with a suitable choice of parameters such multicomponent models can display all the properties associated with simple gradient models (filling in of gaps and formation of duplications with omission of midline structures) or with double-gradient models (symmetry-breaking, replacement of organisers and regulation of proportions). Thus while the evidence for some sort of gradient controlling pattern seems compelling there is no decisive evidence for any particular model, and some of the models are far removed from the popular conception of a 'gradient'.

It is therefore a mistake to expect too close a relationship between experimental and theoretical work. In the present state of the science the theoretical work can show us what is possible and suggest what is reasonable within the time, space and temperature constraints of an embryo. So we know that an explanation for regional specification can be provided by simple chemical kinetics and diffusion. We are not obliged to postulate the involvement of other processes such as vectorial pumping or electrophoretic transport of materials, although we cannot exclude them either. The theoretical work can tell us what is possible but it cannot predict what is actually going on in reality. This must be found out by experiment.

10

Thresholds and repeating patterns

In Chapter 8 we considered reaction systems which could exist in more than one stable steady state and in Chapter 9 we considered smooth variations of cell state with position. The problem of thresholds is the problem of the *emergence* of the discontinuities between regions of the embryo, that is to say the selection, controlled by a chemical signal, of a particular steady state from a set of possibilities. A *threshold* is a line which separates two qualitatively distinct regions. It is sometimes used to refer to geographical boundaries in the embryo, which are lines separating regions of different states of determination, and it is also sometimes used to refer to a line, or surface, in state space which separates two basins of attraction, being then the same as a separatrix.

Switch mechanisms

It is sometimes assumed that the cooperative properties of allosteric enzymes are sufficient to explain the existence of thresholds. But this is not enough. However great the enhancement of binding of substrate to enzyme by positive cooperativity there will still be a smooth relationship between state and position, and this is a gradient and not a threshold. Furthermore, any equilibrium process such as this will be reversible. Thresholds, on the other hand, are not reversible. Once they are formed they persist even in the absence of the interactions which brought about the regionalisation in the first place. Theorists have paid relatively little attention to this problem; of all the models advanced to explain regional specification only a few involve the formation of persistent discontinuities (e.g. Babloyantz & Hiernaux, 1974).

If a threshold is going to be truly discrete and persistent then it must be produced by some sort of switch mechanism. A switch in biochemistry is a bistable reaction system, as discussed in Chapter 8, which is controlled by some sort of intercellular interaction, most simply a concentration gradient of a morphogen. A simple model of this type is depicted in Fig. 10.1 (Lewis, Slack & Wolpert, 1977). Here, a product g is formed from a

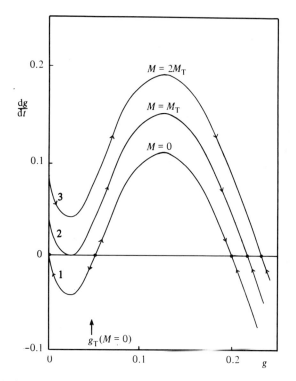

Fig. 10.1. Behaviour of the biochemical switch described in the text. The curves show the net rate of formation plotted against g, the concentration of g. In the absence of the morphogen M ($M = 0$, curve 1) there are three steady states where the curve crosses the abscissa. Above the threshold level g_T, g will spontaneously rise to the upper state and below the threshold it will spontaneously fall to the lowest state. At the critical morphogen concentration M_T (curve 2) the two lower steady states disappear, so for all morphogen concentrations above M_T (e.g. curve 3), g must rise to the upper state.

gene G. Its formation is activated by the morphogen M and also autocatalytically by g itself.

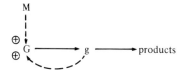

In Fig. 10.1 the rate of synthesis of g is plotted against g, its concentration. The steady states in this system occur where this curve crosses the abscissa, i.e. the points where $dg/dt = 0$. In the absence, or at low concentrations of the morphogen M (curve 1), there are three steady states of which two are stable and one unstable. The stable points are the first and third since for these a slight increase in g leads to a negative dg/dt and vice versa. For the point in the middle a slight increase in g would lead to a positive dg/dt and thus be autocatalytic, and the same is true of a slight decrease. This is therefore unstable.

If the cells start off in the lower stable state and are then exposed to a high morphogen concentration they will follow a trajectory such as curve 3 until they arrive at its sole stable point which corresponds to the upper stable point of curve 1. If the morphogen is now withdrawn the cells are left in the upper state on curve 1 since there is no way that they can now leave it. The switch mechanism therefore has a 'memory' for the transient elevation of the morphogen level. The threshold morphogen concentration M_T corresponds to the critical curve number 2 at which the lower stable point coalesces with the unstable point.

If this mechanism were operating in a field of cells exposed to a gradient of morphogen then all cells with $M > M_T$ would enter and remain in the upper state while all with $M < M_T$ would stay unchanged in the lower state. If the cells were not all identical then the boundary between the two regions would initially be crooked since some cells would respond to $M_T - \delta M$ and others might fail to respond to $M_T + \delta M$. It is popularly supposed that boundaries are later smoothed because the cells of the two regions develop different adhesive properties at their surfaces and a little local movement then brings cells of like type together, but it is not known how generally true this is.

It is not only gradients which need biochemical switches to create sharp boundaries in the tissue. Any mechanism of regional specification entails the existence of biochemical switches. For example, cytoplasmic localisation of a morphogen M will produce a boundary because some cells will inherit a concentration of M above the critical M_T and others will not. A growth zone from which structures emerge in a sequential manner might involve a continuous rise in M in the growing region so that all the cells passed through a sequence of states while they remained in the zone. As cohorts of cells were displaced from it they would remain in the highest state so far attained. This is a possible mechanism for the 'progress zone' model, proposed to explain limb development but often taken to apply to other growth zones as well (Summerbell, Lewis & Wolpert, 1973). When looked at in this way, it is not surprising that part of the body may appear to be formed by a gradient and the remainder by a growth zone mechanism, as is for example found in insect embryos of intermediate-length germ anlage, since at the biochemical level the elements of the mechanisms are the same.

The requirement that the bistable switches underlying threshold formation should have a memory is not affected by the possibility that one of the chemical substances involved is DNA. For example if a stretch of DNA is moved to a new location in the genome then a switch memory is required to prevent it moving back to its old position. There has recently been some interest in models for the stabilisation of decisions based on DNA methylation (Razin & Riggs, 1980). An enzyme called 'maintenance methylase' will methylate positions on a DNA strand opposite positions which are already methylated. The methylation pattern is clonally heritable because every replication of fully methylated DNA produces two molecules consisting of a new unmethylated strand and an old methylated one, both of which will be substrates for the enzyme. There is some evidence that activation of gene activity in terminally differentiated cells is associated with demethylation. If this is confirmed then it implies that inductive signals should produce a transient, specific, demethylation of those genes which are to be activated.

Homeogenetic induction

Biochemical switch mechanisms involving cell surface components can also provide a possible explanation for the phenomenon of homeogenetic induction. This has been investigated mainly for neural induction in the Amphibia, where it has been shown that explants from the neural plate will induce regionally similar structures from gastrula ectoderm either *in vivo* (Mangold, 1933) or *in vitro* (ter Horst, 1948). If a bistable switch mechanism were operative this might be because the switch can be turned on not only by the morphogen M but also by its own product g. If enough of g is added to raise the concentration transiently above the threshold g_T then the autocatalysis will ensure that g is further raised by new synthesis until the upper steady state is reached. Suppose that g is a cell surface molecule and can turn on its own synthesis not only in its own cell but also in the neighbouring cells. It is quite easy to see that the induction could then become self-perpetuating since tissue in which the switch had been turned on could activate another region into which it was grafted. As Meinhardt & Gierer (1980) have emphasised, in normal development there must be mechanisms which limit this sort of infectious spread of determined states, but it remains a possibility in certain experimental situations.

States arising from threshold responses

If an embryonic field becomes partitioned into a number of territories by means of a gradient-and-threshold mechanism then, after the interaction is over, all switches will be 'on' at the original 'high' end and each territory across the field will have one more switch 'off'. If there are five switches and the 'on' state is represented by 1 and the 'off' state by 0, then the states of the territories will be 11111, 01111, 00111, 00011, 00001 and 00000. The same applies to a progress zone mechanism. The subsequent course of development of each territory will then be determined by this combination of switch states or 'epigenetic coding', together with the codings arising from previous decisions.

This sort of mechanism has one major disadvantage from the point of view of those who feel that nature is economical. With serial codings, only $n + 1$ territories can be uniquely named using n switches. However, if the codings could be combinatorial then 2^n could be named with n switches. For example three switches could code for eight territories: 111, 011, 010, 110, 100, 000, 001, 101. If nature really strove for economy then combinatorial epigenetic codings could provide quite a saving. For example, the human body probably contains about $10\,000 \simeq 2^{13}$ named parts. If the primordia were specified with strictly combinatorial codings then 13 switches would suffice for this task, while if each part had at least one unique label then $10\,000$ switches would be required. It is a measure of our ignorance of reality that we do not know which of these alternatives is nearer the truth. The only piece of evidence which bears on the problem suggests that the serial codings really do exist in the *bithorax* system of *Drosphila* (see Chapter 4). When the whole complex is missing from the genome then the late embryo contains ten copies of the mesothorax. If a chromosome segment is introduced which bears the gene Ubx^+ then the nine most posterior copies become converted to metathorax (Lewis, 1978). This suggests that Ubx^+ is not only 'on' in the normal metathorax but in all the abdominal segments as well. The Lewis model presented in Chapter 4 accordingly assumes that the codings of the different regions along the egg axis are serial and not combinatorial.

A combinatorial epigenetic code appears less economical when one considers how it might be set up. An organism consisting of 2^{13} parts could have every part named using only 13 switches but if these states were set up by a succession of binary decisions in response to graded signals then this implies the existence of $2^0 + 2^1 + 2^2 + \ldots + 2^{12} = (2^{13} - 1)$ gradients, which seems rather excessive. It is probably not worth speculating too much about a problem in the almost total absence of evidence. However, if we assume that each switch and each gradient corresponds to at least one gene, then whatever the relative complexity of signals and responses a significant fraction of the genome of higher organisms is likely to consist of genes involved in building the body plan.

Repeating patterns

The segments of an insect larva, or the somites of a vertebrate embryo, are repeating patterns consisting of series of structures which appear similar or identical to one another. The determination of such patterns poses special problems and there are at present two quite different sorts of explanation which have been advanced.

According to the 'positional information' viewpoint there is no essential difference between repeating patterns and any other pattern. Each unit is formed by a separate and independent threshold response to a control variable and so whatever the superficial similarities between the successive territories, their states of determination are fundamentally different or 'non-equivalent'. This view is implicit in a paper by Lawrence (1981) who argues that segments are essentially units of clonal restriction. This may be true operationally but is somewhat unhelpful because clonal restriction does not prove determination (see discussion in Chapters 2, 3 and 4), and because even if insect segments are differently determined at the time of their formation, this in no way contributes to explaining the similarity between them.

According to the alternative viewpoint there is some kind of special mechanism which generates a spatial series of identical determined states, and the recent work on insect development tends to support this view. If the repeating pattern of segments arose solely because of the different states assigned to different regions then the segment boundaries should disappear when two adjacent regions were assigned the same state. However, as we have already seen, the P9 deletion in which the whole *bithorax* complex of *Drosophila* is removed produces not a single giant mesothoracic segment but a serial repetition of mesothoracic segments with a normal total segment number. Conversely there are the other mutants such as *gooseberry*, *patch* and *even-skipped* which alter the repeating pattern and may alter the total segment number but preserve the character of segments appropriate to their position in the body.

This suggests that there are two systems involved in the formation of the body plan: a gradient-and-threshold system to specify the differences along the axis, and another system which superimposes the repeating pattern. To make things more complicated there seems to be an association between the repeating pattern and a gradient *within* each segment. The evidence for this is twofold. Firstly, the mutations of *Drosophila* which disrupt segmentation do not often produce sharp pattern discontinuities. Where segments are apparently fused at non-homologous levels there is usually an interposed region of reversed polarity which joins them together. Secondly, studies on the regeneration of the larval cuticle of the hemimetabolous insect *Oncopeltus* show that when non-homologous levels of adjacent segments are joined there is intercalary regeneration at the junction to restore pattern continuity (Wright & Lawrence, 1981). The behaviour of tissues during regeneration is usually taken to reveal their epigenetic codings and these experiments suggest the existence of a repeating segmental gradient of determination assigned during embryogenesis.

Somitogenesis in vertebrates occurs sequentially from head to tail. Numerous studies have shown that the pre-somite mesoderm need not be continuous for segmentation to occur, in other words the sequence of furrow formation can jump surgical gaps (e.g. Menkes & Sandor, 1977). This implies that the determinative events occur earlier and that no cellular communication or mechanical transmission is necessary for the segmentation process itself. Work on amphibian embryos using temperature shocks has shown an early sensitive phase during gastrulation during which the entire file of somites can be disrupted, and a later sensitive phase which precedes visible segmentation by a few hours and which is experienced by the pre-somite mesoderm at later times for more posterior levels (Pearson & Elsdale, 1979; Elsdale & Pearson, 1979; see Chapter 3). The number of somites in the embryo is not affected by an overall size reduction to half-normal (linear dimensions ×0.8 normal) (Cooke, 1975). This latter result probably holds also for insect segments and chick somites.

So, when we enquire as to what sort of special mechanism might generate repeating structures, it is worth listing the requirements which we place on it: (1) it should produce equivalent structures; (2) it should do so sequentially; (3) it should accommodate regulation of number over a range of perhaps 1.5-fold in linear dimensions; (4) at least in the insect segments it should provide for some sort of intrasegmental gradient which persists through larval life.

Reaction–diffusion models

We have already encountered reaction–diffusion models in connection with symmetry-breaking (Chapter 8) and double gradients (Chapter 9). They have also been advanced as an explanation for repeating patterns (e.g. Turing, 1952; Maynard Smith, 1960). We have seen that there is a critical size of field below which no prepattern can be set up and above this a size range in which a monotonic gradient can form. Above this size things become more complicated since more than one inhomogeneous state may be possible and the one which is reached may depend on the nature of the initial perturbation. However, in general the larger the field the more concentration peaks it is possible to fit on it, and for any given length there will be one prepattern which has the fastest growth rate starting from homogeneity. For example the Gierer–Meinhardt lateral inhibition model can give the multiple peaks shown in Fig. 10.2. The main difference from the figures in the previous chapter in which this model produces monotonic gradients is that the inhibitor diffusion constant has been reduced relative to the size of the field.

This could certainly produce equivalent structures given a single threshold response by the tissue, so that peak and trough regions became differently determined. However, it does not seem to satisfy the other three requirements: there is no reason to expect the units to differentiate

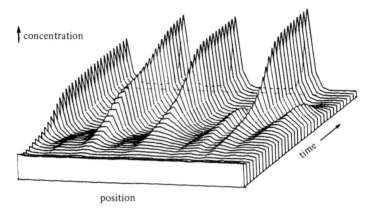

Fig. 10.2. Formation of a periodic pattern in space by the lateral inhibition model of Gierer and Meinhardt. The figure shows the concentration of the inhibitor, which is initially uniform. When inhibitor concentration is subjected to small random fluctuations a number of peaks begin to grow. Some die away and others persist as a stable pattern. (Figure kindly provided by Dr H. Meinhardt.)

sequentially, there is no intrasegmental gradient, and the number of structures will be proportional to the total size of the field. This last point has caused much controversy. It is important to note that although a monotonic gradient can regulate its proportions reasonably well a multiple peak pattern cannot. This is because the natural wavelength of the system is a function of the kinetic and the diffusion constants. Roughly speaking, a given pattern will remain in existence over a change in linear dimensions of up to one wavelength, and this is proportionately a large range for a monotonic gradient but a small range for a multiple peak pattern.

Recently it has been pointed out that this objection could be circumvented if the diffusion constants, which provide the system with an absolute length-scale, were variables which depended in some way on the overall size (Othmer & Pate, 1980; Gierer, 1981). For example a rapidly diffusing substance might be synthesised by all the cells and diffuse out of the field at its margins. Its steady-state level would then increase with total field size and it might regulate the number of gap junctions between cells, which in turn would affect the diffusion constants of the morphogens. The likelihood of such a mechanism is a matter for speculation.

Kauffman, Shymko & Trabert (1978) attempted to use a progressive decline of diffusion constants to explain the early development of *Drosophila*. Their reaction–diffusion model was designed so that only one prepattern or 'mode' could arise in each range of values of the diffusion constants. So the embryo was supposed to pass through a series of modes, starting with a monotonic gradient and progressing to more complex prepatterns fitted onto the ellipsoidal surface of the blastoderm. Each mode triggered a new threshold response and was remembered as a pattern of activation of binary switches. The combinations of switch states which were built up in the embryo accounted for the relative frequencies of transdeterminations between the imaginal discs formed from these regions. This model is most ingenious and could account for segmentation as a response to a prepattern with 16 maxima along the long axis of the egg. However, it is intrinsically incapable of explaining a sequential formation of segments such as is found in most insects and is especially obvious in the short-germ types. It also predicts the formation of twins following a ligation transverse to the long axis, whereas as we saw in Chapter 4 the usual result of this experiment is a gap in the sequence of segments.

Finally, computer simulations of several models of this sort have indicated a rather poor fidelity of the pattern. This has led some authors to suggest that such processes might underlie the formation of imprecise patterns in late development, such as skin pigmentation, but are unlikely to be responsible for the rather precise control of number found for repeating patterns in the general body plan (Bard & Lauder, 1974; Bunow *et al.*, 1980; Murray, 1981; Bard, 1981).

Lateral activation

Another type of reaction–diffusion model which has been advanced to account for repeating patterns depends on two substances whose synthesis is locally mutually exclusive but which activate one another at a distance. This property is possessed by the multicomponent gradient model discussed in the previous chapter, in the case in which there are only two genes involved (Meinhardt & Gierer, 1980). Here g_1 represses g_2 locally because it increases production of the repressor. But s_1 is formed from g_1, diffuses more rapidly and can support the production of g_2 at a distance.

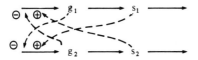

The main difference between this sort of model and the lateral inhibition models is apparent in two dimensions. The lateral inhibition model will generate isolated peaks of activation surrounded by zones of inhibition. By contrast a lateral activation model will lead to the formation of long stripes because the two types of zone are each dependent on the proximity of the other (Fig. 10.3). This property is necessary for the formation of segments because it is clear from the fate maps of insect embryos that the transverse extension of the early segment

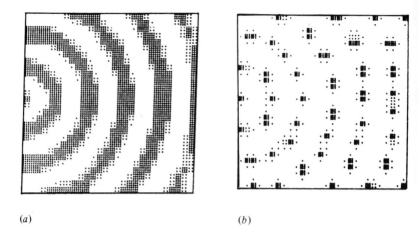

(*a*)   (*b*)

Fig. 10.3. Formation of stripes and spots in a two-dimensional space. (*a*) is a lateral activation model described in the text and (*b*) the lateral inhibition model described in Chapter 9. The development of the patterns was initiated by a disturbance at centre left which becomes the focus of the stripes or spots. (Figure kindly provided by Dr H. Meinhardt.)

rudiments must exceed their longitudinal extension. The model can also add stripes sequentially to one end in a growing or non-growing situation. It cannot, however, accommodate a significant size variation in the field, for the same reason as other reaction–diffusion models: the size of each stripe will be determined by the values of the diffusion constants for g and s, so a small field will accommodate fewer stripes than normal and a large field will acommodate more.

A clock model

It was with the object of accounting both for proportion regulation and for sequential formation of vertebrate somites that the 'clock-and-wave-front' model was advanced (Cooke & Zeeman, 1976). Perhaps it should have been called the 'clock-and-gradient' model to indicate more clearly its two principal components. The gradient is set up early in development and specifies the *rate* at which cells progress towards the act of segmentation. The clock is a biochemical oscillator which operates synchronously throughout the tissue and 'gates' the cells into groups which form each somite. No exact mechanism is advanced for either the gradient or the clock, but the idea is that segmentation requires some change in the state of determination of the cells. All cells will make this change autonomously but those which were at the 'high' end of the gradient will change earlier than those which were at the 'low' end. The clock goes through one cycle

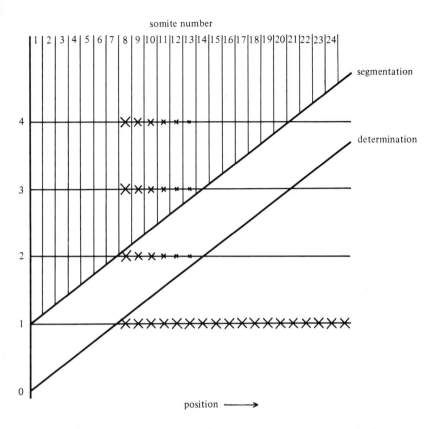

Fig. 10.4. Production of defects in the file of somites by a temperature shock, according to Pearson & Elsdale (1979). A gradient at some earlier stage determines that the tissue shall become segmented from anterior to posterior at progressively later times. The segmentation has two phases: a temperature-sensitive 'determination' and a visible segmentation. At time 1 a temperature shock desynchronises a 'clock' in all the undetermined tissue (**XXXX**). Recovery occurs so that by time 2 the remaining undetermined tissue has become resynchronised, but the tissue which became determined before recovery cannot segment normally. By time 3 this is apparent as a defect in the file of somites, but somites formed subsequently are normal (time 4).

for each segment. It runs synchronously in all cells and advances or retards their progression according to the amplitude of the oscillating concentrations. Hence the cells arrive at the determination event not in a smooth progression but in clumps, and each clump will form one segment.

This accounts for proportion regulation assuming that the original gradient is able to regulate. It accounts for the sequential character of segmentation also by reference to the original gradient. It also accounts for the fact that the segmentation process appears to jump gaps where a

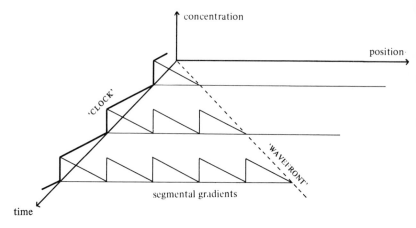

Fig. 10.5. Application of the 'clock-and-wavefront' model to segmentation in insects. The clock produces a sawtooth oscillation of some morphogen which is synchronised in all the cells of the field. The cells become determined in a sequence from left to right. At the instant of determination they 'remember' the morphogen concentration. This converts the sawtooth oscillation into a sawtooth pattern along the egg axis. The 'wavefront' is the level to which determination has proceeded by any given time.

hole is made in advance of the latest-formed segment. In this model long-range interactions are necessary only in the initial stage when the gradient is set up. Short-range interactions may be necessary at the later stage to synchronise the oscillator, but are not necessary for the propagation of the 'wavefronts' of segment determination or the later visible segmentation.

The gradient is presumed to be set up during gastrulation. At this stage brief temperature shocks will produce disturbances which may lie at any position along the later files of somites. During visible somitogenesis, which occurs during neurula and tailbud stages, temperature shocks disturb only those somites which form a few hours later, the exact time depending on species. This may be accounted for if the shock desynchronises the clocks in neighbouring cells and if they then become recoupled after a certain recovery time (Fig. 10.4).

The clock-and-wavefront model can be extended to cover insect segmentation as well. In this case it is necessary not simply to control the number of structures but also to provide a gradient along each segment responsible for subsequent internal regionalisation. We must now assume that the clock is a 'relaxation oscillator' producing a sawtooth variation of concentration with time. We further assume that in this case the determination process results in the memorisation of the clock value at the instant it occurs (Fig. 10.5). This means that a sawtooth concentration variation in time becomes converted into a sawtooth concentration variation in space along the axis of the animal. Such a mechanism could explain the pattern abnormalities produced by many of the *Drosophila* segmentation mutants as resulting from altered waveforms of the oscillator. It cannot, however, easily account for the 'pair rule' mutants in which every other segment is affected, and it also begs questions about how the sawtooth gradient in space could be stabilised after its formation.

The 'clock' models currently seem to account for more of the characteristics of repeating patterns than do the alternative models. However, they have not been spelled out in detail and if they were they might turn out to have defects which are not apparent at present.

11

General conclusions

In this book an attempt has been made to introduce the problem of regional specification in animal development to a wide audience. This has involved firstly trying to define the terminology of experimental embryology in a clear and precise way, secondly presenting the basic experimental data on early developmental decisions, and thirdly introducing some of the models which have been advanced to explain the phenomena.

The experimental section has been summarised in Chapter 7, and it is clear that quite a lot is already known about early development at a cellular level. These results should be taken into account by those who wish to study embryos at the molecular level.

We know from a study of fate maps that embryonic development does not, in general, proceed by random cell differentiation followed by sorting out into a spatial pattern. Rather we find a hierarchy of decisions in each of which cells in different regions of a field become specified to follow one of a certain choice of developmental pathways. Regional commitment can be mediated both by cytoplasmic localisation and by induction, and we suspect that at least in early embryos it occurs first in the cytoplasm and only later in the nucleus. For each level in the hierarchy specification, which is reversible, becomes succeeded by determination, which is not. Each state of determination embodies a competence to make a further set of developmental decisions. Cell movements and cell differentiation are consequences of these decisions.

Homologous inductive interactions occur in different classes of vertebrates and it seems probable that these have common mechanisms. However, it is not known to what extent mechanisms are shared by animals whose embryos have a fundamentally different body plan. Certain responses to experimental interference – such as adaptation of proportions to a change in size, or twinning following division of the embryo – are very widespread but it is not known whether these similarities in behaviour reflect similar biochemistry and similar dynamics, different biochemistry with similar dynamics or merely similar categories in the mind of the investigator.

We know nothing for certain about the biochemistry of regional specification but suspect that the full range of signals and responses may be very complex. We do not even know whether the phenomena can be explained in terms of small subsets of substances or whether a mathematical theory of the 'whole egg' is what is required.

With regard to models the theorists have provided a satisfactory solution to the general problem of symmetry-breaking. They have suggested a new way of looking at developmental decisions in terms of dynamical systems theory, and advanced plausible but unproved models for the determination of repeating and non-repeating series of structures. The formation of mirror-symmetrical patterns and the phenomenon of defect regulation provide strong evidence for the involvement of gradients in regional specification, but a gradient of cell state need not involve physical diffusion of a substance and may involve many substances rather than a single 'morphogen'.

In general the problem with the models is that they cannot really be disproved using present techniques. Even if a model is found wanting in some respect, it can always be elaborated in such a way as to remedy the deficiency. We are left feeling that the models are of more use as an aid to clear thinking than as possible explanations of reality.

It may be that the model-builders' true métier will turn out to be as the embryological architect working in partnership with the genetic engineer, designing the feedback loops and signalling systems for artificial forms of multicellular life. Perhaps it will only be when we have made them ourselves that we shall truly understand how they develop and how they work.

References

Achtelig, M. & Krause, G. (1971). Experimente am ungefurchten Ei von *Pimpla turionellae* L. (Hymenoptera) zur Funktionsanalyse des Oosombereichs. *Wilhelm Roux' Arch. Entwicklungsmech. Org.* **167**, 164–82.

Adelman, H. B. (1930). Experimental studies on development of the eye. III. The effect of the substrate (*Unterlagerung*) on the heterotopic development of median and lateral strips of the anterior end of the neural plate of *Ambystoma. J. Exp. Zool.* **57**, 223–81.

Ancel, P. & Vintemberger, P. (1949). La rotation de symétrisation, facteur de la polarisation dorsoventrale des ébauches primordiales, dans l'oeuf des Amphibiens. *Arch. Anat. Microsc. Morphol. Exp.* **38**, 167–83.

Andronov, A. A., Vitt, A. A. & Khaikain, S. E. (1966). *Theory of Oscillators*. Pergamon Press, Oxford.

Asahi, K., Born, J., Tiedemann, H & Tiedemann, H. (1979). Formation of mesodermal pattern by secondary inducing interactions. *Wilhelm Roux' Arch. Dev. Biol.* **187**, 231–44.

Atkinson, J. W. (1971). Organogenesis in normal and lobeless embryos of the marine prosobranch gastropod *Ilyanassa obsoleta. J. Morphol.* **133**, 339–52.

Azar, Y. & Eyal-Giladi, H. (1979). Marginal zone cells – the primitive streak inducing component of the primary hypoblast in the chick. *J. Embryol. Exp. Morphol.* **52**, 79–88.

Babloyantz, A. (1977). Self organization phenomena resulting from cell–cell contact. *J. Theoret. Biol.* **68**, 551–61.

Babloyantz, A. & Hiernaux, J. (1974). Models for positional information and positional differentiation. *Proc. Natl. Acad. Sci. USA* **71**, 1530–3.

Bäckström, S. (1954). Morphogenetic effects of lithium on the embryonic development of *Xenopus. Ark. Zool.* **6**, 527–36.

Baker, P. C. (1965). Changing serotonin levels in developing *Xenopus laevis. Acta Embryol. Morphol. Exp.* **8**, 197–204.

Balinsky, B. I. (1947). Kinematik des endodermalen Materials bei der Gestaltung der wichstigsten Teile des Darmkanals bei den Amphibien. *Wilhelm Roux' Arch. Entwicklungsmech. Org.* **143**, 126–66.

Balinsky, B. I. (1948). Korrelation in der Entwicklung der Mund und Kiemenregion und des Darmkanals bei Amphibien. *Wilhelm Roux' Arch. Entwicklungsmech. Org.* **143**, 365–95.

Bard, J. B. L. (1981). A model for generating aspects of zebra and other mammalian coat patterns. *J. Theoret. Biol.* **93**, 363–85.

Bard, J. & Lauder, I. (1974). How well does Turing's theory of morphogenesis work? *J. Theoret. Biol.* **45**, 501–31.

Bautzmann, H. (1926). Experimentelle Untersuchungen zur Abgrenzung des Organisationszentrums bei *Triton taeniatus. Wilhelm Roux' Arch. Entwicklungsmech. Org.* **108**, 283–321.

Bautzmann, H., Holtfreter, J., Spemann, H. & Mangold, O. (1932). Versuche zur Analyse der Induktionsmittel in der Embryonalentwicklung. *Naturwissenschaften* **20**, 971–4.

Becker, H. J. (1957). Über Röntgenmosaikflecken und Defektmutationen am Auge von *Drosophila* und die Entwicklungsphysiologie des Auges. *Z. Indukt. Abstamm. Vererbungsl.* **88**, 333–73.

Bellairs, R. (1963). The development of somites in the chick embryo. *J. Embryol. Exp. Morphol.* **11**, 697–714.

Bodmer, W. F. (1981). Gene clusters, genome organization, and complex phenotypes. When the sequence is known what will it all mean? *Am. J. Hum. Genet.* **33**, 664–82.

Boveri, T. (1908). Die Entwicklung dispermer Seeigel-Eier. Ein Beitrag zur Befruchtungslehre und zur Theorie des Kerns. *Z. Naturw. (Jena)* **43**, 1–292.

Boycott, A. E. & Diver, C. (1923). On the inheritance of sinestrality in *Limnaea peregra. Proc. R. Soc. B* **95**, 207–13.

Brachet, J. (1977). An old enigma: the grey crescent of amphibian eggs. *Curr. Top. Dev. Biol.* **11**, 133–86.

Brandhorst, B. P. (1976). Two dimensional gel patterns of protein synthesis before and after fertilization of sea urchin eggs. *Dev. Biol.* **52**, 310–17.

Briggs, R. & King, T. J. (1952). Transplantation of living nuclei from blastula cells into enucleated frogs' eggs. *Proc. Natl. Acad. Sci. USA* **38**, 455–63.

Brinster, R. L. (1974). The effect of cells transferred into the mouse blastocyst on subsequent development. *J. Exp. Med.* **140**, 1049–56.

Buehr, M. & McLaren, A. (1974). Size regulation in chimeric mouse embryos. *J. Embryol. Exp. Morphol.* **31**, 229–34.

Bunow, B., Kernevez, J. P., Joly, G. & Thomas, D. (1980). Pattern formation by reaction-diffusion instabilities: application to morphogenesis in *Drosophila. J. Theoret. Biol.* **84**, 629–49.

Cather, J. N. (1967). Cellular interactions in the development of the shell gland of the gastropod *Ilyanassa. J. Exp. Zool.* **166**, 205–23.

Cather, J. N. & Verdonk, N. H. (1974). The development of *Bithynia tentaculata* (Prosobranchia, Gastropoda) after removal of the polar lobe. *J. Embryol. Exp. Morphol.* **31**, 415–22.

Cather, J. N. & Verdonk, N. H. (1979). Development of *Dentalium* following removal of the D-quadrant blastomeres at successive cleavage stages. *Wilhelm Roux' Arch. Dev. Biol.* **187**, 355–66.

Cavallin, M. (1971). La 'polyembryonie substitutive' et le probleme de l'origine de la lignée germinale chez le Phasme *Carausius morosus* Br. *C.R. Hebd. Séances Acad. Sci., Ser. D* **272**, 462–5.

Chabry, L. (1887). Contribution à l'embryologie normale et tératologie des ascidies simples. *J. Anat. Physiol. (Paris)* **23**, 167–319.

Chan, L. N. & Gehring, W. J. (1971). Determination of blastoderm cells in *Drosophila melanogaster. Proc. Natl. Acad. Sci. USA* **68**, 2217–21.

Chapman, V. M., Adler, D., Labarca, C. & Wudl, L. (1976). Genetic variation of β-glucuronidase expression during early embryogenesis. In *Embryogenesis in Mammals*, CIBA Symposium 40, pp. 115–31. Elsevier, Amsterdam.

Child, C. M. (1941). *Patterns and Problems of Development*. University of Chicago Press.

Chuang, H. H. (1939). Induktionsleistungen von frischen und gekochten Organteilen (Niere, Leber) nach ihrer Verpflanzung in Explantate und verscheidene Wirtsregionen von Tritonkeimen. *Wilhelm Roux' Arch. Entwicklungsmech. Org.* **139**, 556–638.

Clavert, J. (1963). Symmetrisation of the egg of vertebrates. *Adv. Morphogen.* **2**, 27–60.

Clement, A. C. (1952). Experimental studies on germinal localisation in *Ilyanassa*. I. The role of the polar lobe in determination of the cleavage pattern and its influence in later development. *J. Exp. Zool.* **121**, 593–625.

Clement, A. C. (1956). Experimental studies on germinal localisation in *Ilyanassa*. II. The development of isolated blastomeres. *J. Exp. Zool.* **132**, 427–45.

Clement, A. C. (1962). Development of *Ilyanassa* following removal of the D macromere at successive cleavage stages. *J. Exp. Zool.* **149**, 193–215.

Clement, A. C. (1963). Effects of micromere deletion on development in *Ilyanassa*. *Biol. Bull.* **125**, 375 (abstr.).

Clement, A. C. (1967). The embryonic values of the micromeres in *Ilyanassa obsoleta* as determined by deletion experiments. I. The first quartet cells. *J. Exp. Zool.* **166**, 77–88.

Clement, A. C. (1968). Development of the vegetal half of the *Ilyanassa* egg after removal of most of the yolk by centrifugal force compared with the development of animal halves of similar visible composition. *Dev. Biol.* **17**, 165–86.

Cohn, M. & Horibata, K. (1959). Inhibition by glucose of the induced synthesis of the β-galactosidase enzyme system of *Escherichia coli*. Analysis of maintenance. *J. Bacteriol.* **78**, 601–23.

Cole, R. J. & Paul, J. (1965). Properties of cultured preimplantation mouse and rabbit embryos and cell strains derived from them. In *Preimplantation Stages of Pregnancy*, CIBA Foundation Symposium, ed. G. E. W. Wolstenholme & M. O'Connor, pp. 82–122. J. & A. Churchill, London.

Collier, J. R. & McCarthy, M. E. (1981). Regulation of polypeptide synthesis during early embryogenesis of *Ilyanassa obsoleta*. *Differentiation* **19**, 31–46.

Conklin, E. G. (1897). The embryology of *Crepidula*. *J. Morphol.* **13**, 1–226.

Conklin, E. G. (1905*a*). The organization and cell lineage of the ascidian egg. *J. Acad. Nat. Sci. Philadelphia* **13**, 1–119.

Conklin, E. G. (1905*b*). Mosaic development in ascidian eggs. *J. Exp. Zool.* **2**, 145–223.

Cooke, J. (1972). Properties of the primary organization field in the embryo of *Xenopus laevis*. I. Autonomy of cell behaviour at the site of initial organizer formation. *J. Embryol. Exp. Morphol.* **28**, 13–26.

Cooke, J. (1973). Properties of the primary organization field in the embryo of *Xenopus laevis*. V. Regulation after removal of the head organizer in normal early gastrulae and in those already possessing a second implanted organizer. *J. Embryol. Exp. Morphol.* **30**, 283–300.

Cooke, J. (1975). Control of somite number during morphogenesis of a vertebrate, *Xenopus laevis*. *Nature (Lond.)* **254**, 196–9.

Cooke, J. (1979). Cell number in relation to primary pattern formation in the embryo of *Xenopus laevis*. I. The cell cycle during new pattern formation in response to implanted organisers. *J. Embryol. Exp. Morphol.* **51**, 165–82.

Cooke, J. (1981). Scale of body pattern adjusts to available cell number in amphibian embryos. *Nature (Lond.)* **290**, 775–8.

Cooke, J. & Zeeman, E. C. (1976). A clock and wave-front model for control of the number of repeated structures during animal morphogenesis. *J. Theoret. Biol.* **58**, 455–76.

Crampton, H. E. (1896). Experimental studies on gasteropod development. *Arch. Entwicklungsmech. Org.* **3**, 1–26.

Crick, F. (1970). Diffusion in embryogenesis. *Nature (Lond.)* **225**, 420–2.

Crick, F. H. C. & Lawrence, P. A. (1975). Compartments and polyclones in insect development. *Science* **189**, 340–7.

Dalcq, A. (1932). Etude des localisations germinales dans l'oeuf vierge d'ascidie par des expériences de mérogonie. *Arch. Anat. Microsc. Morphol. Exp.* **28**, 223–333.

Dalcq, A. & Pasteels, J. (1937). Une conception nouvelle des bases physiologiques de la morphogénèse. *Arch. Biol.* **48**, 669–710.

Dan-Sohkawa, M. & Satoh, N. (1978). Studies on dwarf larvae developed from isolated blastomeres of the starfish *Asterina pectinifera*. *J. Embryol. Exp. Morphol.* **46**, 171–85.

Davidson, E. H. (1976). *Gene Activity in Early Development*, 2nd edn. Academic Press, New York & London.

Davidson, E. H. & Britten, R. J. (1979). Regulation of gene expression: possible role of repetitive sequences. *Science* **204**, 1052–9.

Davidson, E. H., Hough-Evans, B. R. & Britten, R. J. (1982). Molecular biology of the sea urchin embryo. *Science* **217**, 17–26.

Deol. M. S. & Whitten, W. K. (1972). Time of X chromosome inactivation in retinal melanocytes of the mouse. *Nature New Biol.* **238**, 159–60.

Deppe, U., Schierenberg, E., Cole, T., Krieg, C., Schmitt, D., Yoder, B. & von Ehrenstein, G. (1978). Cell lineages of the embryo of the nematode *Caenorhabditis elegans. Proc. Natl. Acad. Sci. USA* **75**, 376–80.

Deuchar, E. M. & Burgess, A. M. C. (1967). Somite segmentation in amphibian embryos: is there a transmitted control mechanism? *J. Embryol. Exp. Morphol.* **17**, 349–58.

Dewey, M. J. & Mintz, B. (1980). Teratocarcinoma cells as agents for producing mutant mice. In *Differentiation and Neoplasia*, pp. 275–82. Springer Verlag, Berlin & Heidelberg.

Diwan, S. B. & Stevens, L. C. (1976). Development of teratomas from the ectoderm of mouse egg cylinders. *J. Natl. Cancer Inst.* **57**, 937–42.

Dohmen, M. R. & Verdonk, N. H. (1979). The ultrastructure and role of the polar lobe in development of molluscs. In *Determinants of Spatial Organization*, ed. S. Subtelny & I. R. Konigsberg, pp. 3–27. Academic Press, New York & London.

Driesch, H. (1891). Entwicklungsmechanische Studien. I. Der Werth der beiden ersten Furchungszellen in der Echinodermenentwicklung. Experimentelle Erzeugung von Theil und Doppelbildungen. *Z. Wiss. Zool.* **53**, 160–78.

Driesch, H. & Morgan, T. H. (1895). Zur Analyse der ersten Entwicklungsstadien des Ctenophoreneies. I. Von der Entwicklung einzelner Ctenophorenblastomeren. *Arch. Entwicklungsmech. Org.* **2**, 204–15.

Ducibella, T. & Anderson, E. (1975). Cell shape and membrane changes in the eight cell mouse embryo: prerequisites for morphogenesis of the blastocyst. *Dev. Biol.* **47**, 45–58.

Dunis, D. A. & Namenwirth, M. (1977). The role of grafted skin in the regeneration of X-irradiated axolotl limbs. *Dev. Biol.* **56**, 97–109.

Dziadek, M. (1979). Cell differentiation in isolated inner cell masses of mouse blastocysts *in vitro*: onset of specific gene expression. *J. Embryol. Exp. Morphol.* **53**, 367–79.

Edelstein, B. B. (1972). The dynamics of cellular differentiation and associated pattern formation. *J. Theoret. Biol.* **37**, 221–43.

Elsdale, T. & Jones, J. (1963). The independence and interdependence of cells in the amphibian embryo. *Symp. Soc. Exp. Biol.* **17**, 257–73.

Elsdale, T. & Pearson, M. (1979). Somitogenesis in amphibia. II. Origins in early embryogenesis of two factors involved in somite specification. *J. Embryol. Exp. Morphol.* **53**, 245–67.

Eppig, J. J., Kozak, L. P., Eicher, E. M. & Stevens, L. C. (1977). Ovarian teratomas in mice are derived from oocytes that have completed their first meiotic division. *Nature (Lond.)* **269**, 517–18.

Evans, M. (1981). Origin of mouse embryonal carcinoma cells and the possibility of their direct isolation into tissue culture. *J. Reprod. Fert.* **62**, 625–31.

Evans, M. J. & Kaufman, M. H. (1981). Establishment in culture of pluripotential cells from mouse embryos. *Nature (Lond.)* **292**, 154–6.

Fankhauser, G. (1945). The effects of changes in chromosome number on amphibian development. *Q. Rev. Biol.* **20**, 20–78.

Farfaglio, G. (1963). Experiments on the formation of the ciliated plates in ctenophores. *Acta Embryol. Morphol. Exp.* **6**, 191–203.

Feldman, J., Gilula, N. B. & Pitts, J. D. (1978). *Intercellular Junctions and Synapses.* Receptors and Recognition, Series B No. 2. Chapman & Hall, London.

Ferrus, A. & Kankel, D. R. (1981). Cell lineage relationships in *Drosophila melanogaster*: the relationships of cuticular to internal tissues. *Dev. Biol.* **85**, 485–504.

Fielding, C. J. (1967). Developmental genetics of the mutant *grandchildless* of *Drosphilia subobscura. J. Embryol. Exp. Morphol.* **17**, 375–84.

Freeman, G. (1976a). The role of cleavage in the localization of developmental potential in the ctenophore *Mnemiopsis leidyi. Dev. Biol.* **49**, 143–77.

Freeman, G. (1976*b*). The effect of altering the position of cleavage planes on the process of localization of developmental potential in ctenophores. *Dev. Biol.* **51**, 332–7.

Freeman, G. (1977). The establishment of the oral–aboral axis in the ctenophore embryo. *J. Embryol. Exp. Morphol.* **42**, 237–60.

Freeman, G. & Lundelius, J. W. (1982). The developmental genetics of dextrality and sinestrality in the gastropod *Limnaea peregra*. *Wilhelm Roux' Arch. Dev. Biol.* **191**, 69–83.

Freeman, G. & Reynolds, G. T. (1973). The development of bioluminescence in the ctenophore *Mnemiopsis leidyi*. *Dev. Biol.* **31**, 61–100.

Galau, G. A., Britten, R. J. & Davidson, E. H. (1974). A measurement of the sequence complexity of polysomal messenger RNA in sea urchin embryos. *Cell* **2**, 9–20.

Galau, G. A., Klein, W. H., Davis, M. M., Wold, B. J., Britten, R. J. & Davidson, E. H. (1976). Structural gene sets active in embryos and in adult tissues of the sea urchin. *Cell* **7**, 487–505.

Gallera, J. (1971). Primary induction in birds. *Adv. Morphogen.* **9**, 149–80.

Gallera, J. & Nicolet, G. (1969). Le pouvoir inducteur de l'endoblaste présomptif contenu dans la ligne primitive jeune de poulet. *J. Embryol. Exp. Morphol.* **21**, 105–18.

Garcia-Bellido, A. & Merriam, J. R. (1969). Cell lineage of the imaginal discs in *Drosophila* gynandromorphs. *J. Exp. Zool.* **170**, 61–76.

Gardner, R. L. (1972). An investigation of inner cell mass and trophoblast tissues following their isolation from the mouse blastocysts. *J. Embryol. Exp. Morphol.* **28**, 279–312.

Gardner, R. L. (1978). The relationship between cell lineage and differentiation in the early mouse embryo. In *Genetic Mosaics and Cell Differentiation*, ed. W. J. Gehring, pp. 205–41. Springer Verlag, Berlin & Heidelberg.

Gardner, R. L. (1982). Investigation of cell lineage and differentiation in the extraembryonic endoderm of the mouse embryo. *J. Embryol. Exp. Morphol.* **68**, 175–98.

Gardner, R. L. & Johnson, M. H. (1973). Investigation of early mammalian development using interspecific chimeras between rat and mouse. *Nature New Biol.* **246**, 86–9.

Gardner, R. L. & Papaioannou, V. E. (1975). Differentiation in the trophectoderm and inner cell mass. In *The Early Development of Mammals*, BSDB Symposium 2, ed. M. Balls & A. E. Wild, pp. 107–32. London: Cambridge University Press.

Gardner, R. L., Papaioannou, V. E. & Barton, S. C. (1973). Origin of the ectoplacental cone and secondary giant cells in mouse blastocysts reconstituted from isolated trophoblast and inner cell mass. *J. Embryol. Exp. Morphol.* **30**, 561–72.

Gardner, R. L. & Rossant, J. (1979). Investigation of the fate of 4.5 day *post coitum* mouse inner cell mass cells by blastocyst injection. *J. Embryol. Exp. Morphol.* **52**, 141–52.

Garner, W. & McLaren, A. (1974). Cell distribution in chimeric mouse embryos before implantation. *J. Embryol. Exp. Morphol.* **32**, 495–503.

Gerhart, J., Ubbels, G., Black, S., Hara, K. & Kirschner, M. (1981). A reinvestigation of the role of the grey crescent in axis formation in *Xenopus laevis*. *Nature (Lond.)* **292**, 511–16.

Gierer, A. (1981). Some physical, mathematical and evolutionary aspects of biological pattern formation. *Phil. Trans. R. Soc. Lond. B* **295**, 429–40.

Gierer, A. & Meinhardt, H. (1972). A theory of biological pattern formation. *Kybernetik* **12**, 30–9.

Glade, R. W., Burrill, E. M. & Falk, R. J. (1967). The influence of a temperature gradient on bilateral symmetry in *Rana pipiens*. *Growth* **31**, 231–49.

Grabowski, C. T. (1957). The induction of secondary embryos in the early chick blastoderm by grafts of Hensen's node. *Am. J. Anat.* **101**, 101–27.

Grobstein, C. (1967). Mechanisms of organogenetic tissue interactions. *Natnl. Cancer Inst. Monogr.* **26**, 279–99.

Gross, K. W., Jacobs-Lorena, M., Baglioni, C. & Gross, P. R. (1973). Cell free translation of maternal messenger RNA from sea urchin eggs. *Proc. Natl. Acad. Sci. USA* **70**, 2614–18.

Gross, P. R. & Coustineau, G. H. (1964). Macromolecule synthesis and the influence of actinomycin on early development. *Exp. Cell Res.* **33**, 368–95.

Guerrier, P. (1970*a*). Les caractères de la segmentation et la détermination de la polarité dorsoventrale dans le développement de quelques Spiralia. I. Les formes à premier clivage égal. *J. Embryol. Exp. Morphol.* **23**, 611–37.

Guerrier, P. (1970*b*). Les caractères de la segmentation et la détermination de la polarité dorsoventrale dans le développement de quelques Spiralia. II. *Sabellaria alveolata* (Annélide polychète). *J. Embryol. Exp. Morphol.* **23**, 639–65.

Guerrier, P. (1970*c*). Les caractères de la segmentation et la détermination de la polarité dorsoventrale dans le développement de quelques Spiralia. III. *Pholas dactylus* et *Spisula subtruncata* (Mollusques lamellibranches). *J. Embryol. Exp. Morphol.* **23**, 667–92.

Gurdon, J. B. (1974). *The Control of Gene Expression in Animal Development.* Oxford University Press.

Hamilton, L. (1969). The formation of somites in *Xenopus. J. Embryol. Exp. Morphol.* **22**, 253–64.

Hamilton, W. J. & Mossman, H. W. (1976). *Human Embryology.* Macmillan, London.

Handyside, A. H. (1978). Time of commitment of inside cells isolated from preimplantation mouse embryos. *J. Embryol. Exp. Morphol.* **45**, 37–53.

Handyside, A. H. (1980). Distribution of antibody and lectin binding sites on dissociated blastomeres from mouse morulae: evidence for polarization at compaction. *J. Embryol. Exp. Morphol.* **60**, 99–116.

Harvey, E. B. (1933). Development of the parts of sea urchin eggs separated by centrifugal force. *Biol. Bull.* **64**, 125–48.

Hillman, N., Sherman, M. I. & Graham, C. (1972). The effect of spatial arrangement on cell determination during mouse development. *J. Embryol. Exp. Morphol.* **28**, 263–78.

Hirose, G. & Jacobson, M. (1979). Clonal organization of the central nervous system of the frog. I. Clones stemming from individual blastomeres of the 16 cell and earlier stages. *Dev. Biol.* **71**, 191–202.

Hogan, B. L. M., Cooper, A. R. & Kurkinen, M. (1980). Incorporation into Reichert's membrane of laminin-like extracellular proteins synthesized by parietal endoderm cells of the mouse embryo. *Dev. Biol.* **80**, 289–300.

Hogan, B. L. M., Taylor, A. & Adamson, E. (1981). Cell interactions modulate embryonal carcinoma cell differentiation into parietal or visceral endoderm. *Nature (Lond.)* **291**, 235–7.

Hogan, B. & Tilly, R. (1978*a*). *In vitro* development of inner cell masses isolated immunosurgically from mouse blastocysts. I. Inner cell masses from 3.5 day p.c. blastocysts incubated for 24 h before immunosurgery. *J. Embryol. Exp. Morphol.* **45**, 93–105.

Hogan, B. & Tilly, R. (1978*b*). *In vitro* development of inner cell masses isolated immunosurgically from mouse blastocysts. II. Inner cell masses from 3.5 to 4.0 day p.c. blastocysts. *J. Embryol. Exp. Morphol.* **45**, 107–21.

Hogan, B. L. M. & Tilly, R. (1981). Cell interactions and endoderm differentiation in cultured mouse embryos. *J. Embryol. Exp. Morphol.* **62**, 379–91.

Holder, N. (1978). The onset of osteogenesis in the developing chick limb. *J. Embryol. Exp. Morphol.* **44**, 15–29.

Holtfreter, J. (1933). Organisierungsstufen nach regionaler Kombination von Entomesoderm mit Ektoderm. *Biol. Zentralbl.* **53**, 404–31.

Holtfreter, J. (1936). Regionale Induktionen in xenoplastisch zusammengesetzten Explantaten. *Wilhelm Roux' Arch. Entwicklungsmech. Org.* **134**, 466–550.

Holtfreter, J. (1938*a*). Differenzierungspotenzen isolierter Teile der Urodelengastrula. *Wilhelm Roux' Arch. Entwicklungsmech. Org.* **138**, 522–656.

Holtfreter, J. (1938*b*). Differenzierungspotenzen isolierter Teile der Anurengastrula. *Wilhelm Roux' Arch. Entwicklungsmech. Org.* **138**, 657–738.

Holtfreter-Ban, H. (1965). Differentiation capacities of Spemann's organizer investigated in explants of diminishing size. PhD thesis, University of Rochester, New York.

Holtzer, H. (1978). Cell lineages, stem cells and the 'quantal' cell cycle concept. In *Stem Cells and Tissue Homeostasis*, BSCB Symposium 2, ed. B. I. Lord, C. S. Potten & R. J. Cole, pp. 1–27. Cambridge University Press.

Hornbruch, A., Summerbell, D. & Wolpert, L. (1979). Somite formation in the early chick embryo following grafts of Hensen's node. *J. Embryol. Exp. Morphol.* **51**, 51–62.

Horst, J. ter (1948). Differenzierungs und Induktionsleistungen verschiedener Abschnitte der Medullarplatte und der Urdarmdaches von *Triton* im Kombinat. *Wilhelm Roux' Arch. Entwicklungsmech. Org.* **143**, 275–303.

Hörstadius, S. (1935). Über die Determination im Verlaufe der Eiachse bei Seeigeln. *Publ. Staz. Zool. Napoli* **14**, 251–429.

Hörstadius, S. (1939). The mechanics of sea urchin development studied by operative methods. *Biol. Rev.* **14**, 132–79.

Hörstadius, S. (1957). On the regulation of bilateral symmetry in plutei with exchanged meridional halves and in giant plutei *J. Embryol. Exp. Morphol.* **5**, 60–73.

Hörstadius, S. (1973). *The Experimental Embryology of Echinoderms.* Clarendon Press, Oxford.

Hotta, Y. & Benzer, S. (1972). Mapping of behaviour on *Drosophila* mosaics. *Nature (Lond.)* **240**, 527–35.

Hsu, Y. C. (1980). Embryo growth and differentiation factors in embryonic sera of mammals. *Dev. Biol.* **76**, 465–74.

Hsu, Y. C. & Gonda, M. A. (1980) Monozygotic twin formation in mouse embryos *in vitro*. *Science* **209**, 605–6.

Hutchins, R. & Brandhorst, B. P. (1979). Commitment to vegetalized development in sea urchin embryos: failure to detect changes in patterns of protein synthesis. *Wilhelm Roux' Arch. Dev. Biol.* **186**, 95–102.

Huxley, J. S. & de Beer, G. A. (1934). *The Elements of Experimental Embryology.* Cambridge University Press. (Reprinted by Hafner, New York, in 1963.)

Illmensee, K. (1976). Nuclear and cytoplasmic transfer in *Drosophila*. In *Insect Development*, ed. P. A. Lawrence, pp. 76–96. Blackwell Scientific, Oxford.

Illmensee, K. (1978). Reversion of malignancy and normalised differentiation of teratocarcinoma cells in chimeric mice. In *Gatlinberg Symposium on Genetic Mosaics and Chimeras in Mammals*, ed. L. B. Russell, pp. 3–25. Plenum Press, New York.

Illmensee, K. & Hoppe, P. C. (1981). Nuclear transplantation in *Mus musculus*: developmental potential of nuclei from preimplantation embryos. *Cell* **23**, 9–18.

Illmensee, K. & Mahowald, A. P. (1974). Transplantation of posterior polar plasm in *Drosophila*. Induction of germ cells at the anterior pole of the egg. *Proc. Natl. Acad. Sci. USA* **71**, 1016–20.

Illmensee, K. & Mintz, B. (1976). Totipotency and normal differentiation of single teratocarcinoma cells cloned by injection into blastocysts. *Proc. Natl. Acad. Sci. USA* **73**, 549–53.

Jacobson, M. & Hirose, G. (1981). Clonal organization of the central nervous system of the frog. *J. Neurosci.* **1**, 271–84.

Janning, W. (1978). Gynandromorph fate maps in *Drosophila*. In *Genetic Mosaics and Cell Differentiation*, ed. W. J. Gehring, pp. 1–28. Springer Verlag, Berlin & Heidelberg.

Johnson, M. H. & Ziomek, C. A. (1981). The foundation of two distinct cell lineages within the mouse morula. *Cell* **24**, 71–80.

Judd, B. H. & Young, M. W. (1973). An examination of the one cistron: one chromomere concept. *Cold Spring Harbor Symp. Quant. Biol.* **38**, 573–9.

Jung, E. (1966). Untersuchungen am Ei des Speisebohnenkäfers *Bruchidius obtectus* SAY (Coleoptera). II. Entwicklungsphysiologische Ergebnisse der Schnürungsexperimente. *Wilhelm Roux' Arch. Entwicklungsmech. Org.* **157**, 320–92.

Kalthoff, J. (1971). Position of targets and period of competence for UV induction of the malformation 'double abdomen' in the egg of *Smittia* sp. *Wilhelm Roux' Arch. Entwicklungsmech. Org.* **168**, 63–84.

Kalthoff, K. (1973). Action spectra for UV induction and photoreversal of a switch in the developmental program of an insect (*Smittia*). *Photochem. Photobiol.* **18**, 355–64.

Kalthoff, K. & Sander, K. (1968). Die Entwicklung der Missbildung 'Doppelabdomen' im partiell UV-bestrahlten Ei von *Smittia parthenogenetica* (Dipt. Chironomidae). *Wilhelm Roux' Arch. Entwicklungsmech. Org.* **161**, 129–46.

Kaneda, T. & Hama, T. (1979). Studies on the formation and state of determination of the trunk organiser in the newt *C. pyrrhogaster*. *Wilhelm Roux' Arch. Dev. Biol.* **187**, 25–34.

Kauffman, S. A. (1971). Gene regulation networks: a theory for their global structure and behaviours. *Curr. Top. Dev. Biol.* **6**, 145–82.

Kauffman, S. A. (1975). Control circuits for determination and transdetermination: interpreting positional information in a binary epigenetic code. In *Cell Patterning*, CIBA Symposium 29 (new series), pp. 201–21. Elsevier, Amsterdam.

Kauffman, S. A., Shymko, R. M. & Trabert, K. (1978). Control of sequential compartment formation in *Drosophila*. *Science* **199**, 259–70.

Kaufman, M. H. & O'Shea, K. S. (1978). Induction of monozygotic twinning in the mouse. *Nature (Lond.)* **276**, 707–8.

Keller, R. E. (1975). Vital dye mapping of the gastrula and neurula of *Xenopus laevis*. I. Prospective areas and morphogenetic movements of the superficial layer. *Dev. Biol.* **42**, 222–41.

Keller, R. E. (1976). Vital dye mapping of the gastrula and neurula of *Xenopus laevis*. II. Prospective areas and morphogenetic movements of the deep layer. *Dev. Biol.* **51**, 118–37.

Kelly, S. J. (1977). Studies of the developmental potential of 4 and 8 cell stage mouse blastomeres. *J. Exp. Zool.* **200**, 365–76.

Kimble, J. E. (1981). Strategies for control of pattern formation in *Caenorhabditis elegans*. *Phil. Trans. R. Soc. B* **295**, 539–51.

Kirschner, M., Gerhart, J. C., Hara, K. & Ubbels, G. A. (1980). Initiation of the cell cycle and establishment of bilateral symmetry in *Xenopus* eggs. *Symp. Soc. Dev. Biol.* **38**, 187–215.

Kleinsmith, L. J. & Pierce, G. B. (1964). Multipotentiality of single embryonal carcinoma cells. *Cancer Res.* **24**, 1544–51.

Kobayakawa, Y. & Kubota, H. Y. (1981). Temporal pattern of cleavage and the onset of gastrulation in amphibian embryos developed from eggs with reduced cytoplasm. *J. Embryol. Exp. Morphol.* **62**, 83–94.

Krause, G. (1939). Die Eitypen der Insekten. *Biol. Zentralbl.* **59**, 495–536.

Krause, G. (1958). Induktionssysteme in der Embryonalentwicklung von Insekten. *Ergebn. Biol.* **20**, 159–98.

Krause, G. & Krause, J. (1965). Über das Vermögen median durchschnittener Keimanlagen von *Bombyx mori* L. Sich *in ovo* und sich ohne Dottersystem *in vitro* zwillingsartig zu entwickeln. *Z. Naturforsch. B* **20**, 334–9.

Lacalli, T. C. & Harrison, L. G. (1978). The regulatory capacity of Turing's model for morphogenesis, with application to slime moulds. *J. Theoret. Biol.* **70**, 273–95.

Lallier, R. (1975). Animalization and vegetalization. In *The Sea Urchin Embryo*, ed. G. Czihak, pp. 473–509. Springer Verlag, Berlin & Heidelberg.

La Spina, R. (1958). Lo spostamento dei plasmi mediante centrifugazione nell'uovo vergine di ascidie e il consequente sviluppo. *Acta Embryol. Morphol. Exp.* **2**, 66–78.

Laufer, J. L., Bazzicalupo, P. & Wood, W. B. (1980). Segregation of developmental potential in early embryos of *Caenorhabditis elegans*. *Cell* **10**, 569–77.

Lawrence, P. A. (1973). A clonal analysis of segment development in *Oncopeltus* (Hemiptera). *J. Embryol. Exp. Morphol.* **30**, 681–99.

228   References

Lawrence, P. A. (1981). The cellular basis of segmentation in insects. *Cell* **26**, 3–10.

Lawrence, P. A. & Morata, G. (1977). The early development of mesothoracic compartments in *Drosophila*. An analysis of cell lineage and fate mapping and an assessment of methods. *Dev. Biol.* **56**, 40–51.

Leblond, C. P. (1972). Growth and renewal. In *Regulation of Organ and Tissue Growth*, ed. R. J. Goss, pp. 13–39. Academic Press, New York & London.

Le Douarin, N. (1971). Caractèristiques ultrastructurales du noyau interphasique chez la caille et chez le poulet et utilization de cellules de caille comme 'marqueurs biologiques' en embryologie expérimentale. *Ann. Embryol. Morphogenet.* **4**, 125–35.

Lehmann, F. E. (1926). Entwicklungsstörungen in der Medullaranlage von *Triton*, erzeugt durch Unterlagerungsdefekte. *Wilhelm Roux' Arch. Entwicklungsmech. Org.* **108**, 243–82.

Lehmann, F. E. (1937). Mesodermisierung des praesumptiven Chordamaterials durch Einwirkung von Lithiumchlorid auf die Gastrula von *Triturus alpestris*. *Wilhelm Roux' Arch. Entwicklungsmech. Org.* **136**, 112–46.

Lewin, B. (1980). Genes and gene number. In *Gene Expression*, vol. 2, 2nd edn, pp. 479–502. Wiley, New York.

Lewis, E. B. (1963). Genes and developmental pathways. *Am. Zool.* **3**, 33–56.

Lewis, E. B. (1978). A gene complex controlling segmentation in *Drosophila*. *Nature (Lond.)* **276**, 565–70.

Lewis, J., Slack, J. M. W. & Wolpert, L. (1977). Thresholds in development. *J. Theoret. Biol.* **65**, 579–90.

Lewis, J. H. & Wolpert, L. (1976). The principle of non-equivalence in development. *J. Theoret. Biol.* **62**, 479–90.

Lillie, F. R. (1906). Observations and experiments concerning the elementary phenomena of embryonic development in *Chaetopterus*. *J. Exp. Zool.* **3**, 153–268.

Lo, C. W. & Gilula, N. B. (1979). Gap junctional communication in the preimplantation mouse embryo. *Cell* **18**, 399–409.

Lohs-Schardin, M. (1982). *Dicephalic* – a *Drosophila* mutant affecting polarity in follicle organization and embryonic patterning. *Wilhelm Roux' Arch. Dev. Biol.* **191**, 28–36.

Lohs-Schardin, M., Cremer, Ch. & Nüsslein-Volhard, Ch. (1979). A fate map for the larval epidermis of *Drosophila melanogaster*. Localised cuticular defects following irradiation of the blastoderm with an ultraviolet laser microbeam. *Dev. Biol.* **73**, 239–55.

Lutz, H. (1949). Sur la production expérimentale de la polyembryonie et de la monstruosité double chez les oiseaux. *Arch. Anat. Microsc. Morphol. Exp.* **38**, 79–144.

Lyon, E. P. (1906). Some results of centrifugalizing the eggs of *Arbacia*. *Am. J. Physiol.* **15**, xxi–xxii (abstr.).

McKinnell, R. G. (1978). *Cloning, Nuclear Transplantation in Amphibia*. University of Minnesota Press, Minneapolis.

McLaren, A. (1972). Numerology of development. *Nature (Lond.)* **239**, 274–6.

McLaren, A. (1976). *Mammalian Chimaeras*. Cambridge University Press.

McMahon, D. (1973). A cell contact model for position determination in development. *Proc. Natl. Acad. Sci. USA* **70**, 2396–400.

Madhavan, M. M. & Schneiderman, H. A. (1977). Histological analysis of the dynamics of growth of imaginal discs and histoblast nests during the larval development of *Drosophila melanogaster*. *Wilhelm Roux' Arch. Dev. Biol.* **183**, 269–305.

Malacinski, G. M., Benford, H. & Chung, H. M. (1975). Association of an ultraviolet irradiation sensitive cytoplasmic localization with the future dorsal side of the amphibian egg. *J. Exp. Zool.* **191**, 97–110.

Malacinski, G. M. & Chung, H. M. (1981). Establishment of the site of involution at novel locations on the amphibian embryo. *J. Morphol.* **169**, 149–59.

Mangold, O. (1933). Über die Induktionsfähigkeit der verscheidener Bezirke der Neurula von Urodelen. *Naturwissenschaften* **21**, 761–6.

Mangold, O. & Seidel, F. (1927). Homoplastische und heteroplastische Verschmelzung ganzer Tritonkeime. *Wilhelm Roux' Arch. Entwicklungsmech. Org.* **111**, 593–665.

Marcus, N. H. (1979). Developmental aberrations associated with twinning in laboratory reared sea urchins. *Dev. Biol.* **70**, 274–7.

Martin, G. R. (1981). Isolation of a pluripotent cell line from early mouse embryos cultured in a medium conditioned by teratocarcinoma stem cells. *Proc. Natl. Acad. Sci. USA* **78**, 7634–8.

Martin, G. R. & Evans, M. J. (1975). Differentiation of clonal lines of teratocarcinoma cells: formation of embryoid bodies *in vitro*. *Proc. Natl. Acad. Sci. USA* **72**, 1441–5.

Masui, Y. (1961). Mesodermal and endodermal differentiation of the presumptive ectoderm of *Triturus* gastrulae through the influence of lithium ion. *Experientia* **17**, 458–9.

Mayer, B. (1935). Über das Regulations- und Induktionsvermögen der halbseitigen oberen Urmundlippe von *Triton*. *Wilhelm Roux' Arch. Entwicklungsmech. Org.* **133**, 518–81.

Maynard-Smith, J. (1960). Continuous, quantised and modal variation. *Proc. R. Soc. B* **152**, 397–409.

Meinhardt, H. (1977). A model of pattern formation in insect embryogenesis. *J. Cell. Sci.* **23**, 117–39.

Meinhardt, H. & Gierer, A. (1974). Applications of a theory of biological pattern formation based on lateral inhibition. *J. Cell Sci.* **15**, 321–46.

Meinhardt, H. & Gierer, A. (1980). Generation and regeneration of sequences of structures during morphogenesis. *J. Theoret. Biol.* **85**, 429–50.

Menkes, B. & Sandor, S. (1977). Somitogenesis: regulation, potencies, sequence determination and primordial interactions. In *Vertebrate Limb and Somite Development*, BSDB Symposium 3, ed. D. A. Ede, J. R. Hinchliffe & M. Balls, pp. 403–19. Cambridge University Press.

Merriam, J. R. (1978). Estimating primordial cell numbers in *Drosophila* imaginal discs and histoblasts. In *Genetic Mosaics and Cell Differentiation*, ed. W. J. Gehring, pp. 71–96. Springer Verlag, Berlin & Heidelberg.

Messenger, E. A. & Warner, A. E. (1979). The function of the sodium pump during differentiation of amphibian embryonic neurones. *J. Physiol. (Lond.)* **292**, 85–105.

Minganti, A. (1949). Transplantations d'un fragment du territoire somitique présomptif de la jeune gastrula chez l'Axolotl et chez le Triton. *Arch. Biol. (Paris)* **61**, 251–355.

Mintz, B. (1965). Experimental genetic mosaicism in the mouse. In *Preimplantation Stages of Pregnancy*, CIBA Foundation Symposium, ed. G. E. W. Wolstenholme & M. O'Connor, pp. 194–216. J. & A. Churchill, London.

Mintz, B. (1970). Gene expression in allophenic mice. In *Control Mechanisms in the Expression of Cellular Phenotypes*, ed. H. A. Padykula, pp. 15–42. Academic Press, New York & London.

Mintz, B. & Illmensee, K. (1975). Normal genetically mosaic mice produced from malignant teratocarcinoma cells. *Proc. Natl. Acad. Sci. USA* **72**, 3585–9.

Monk, M. & Harper, M. I. (1979). Sequential X chromosome inactivation coupled with cellular differentiation in early mouse embryos. *Nature (Lond.)* **281**, 311–13.

Morgan, T. H. (1901). *Regeneration*. Macmillan, London.

Morgan, T. H. (1927). *Experimental Embryology*. Columbia University Press, New York.

Munro, M. & Crick, F. H. C. (1971). The time needed to set up a gradient: detailed calculations. *Symp. Soc. Exp. Biol.* **25**, 439–53.

Murray, J. D. (1981). A prepattern formation mechanism for animal coat markings. *J. Theoret. Biol.* **88**, 161–99.

Nakamura, O. & Matsuzawa, T. (1967). Differentiation capacity of the marginal zone in the morula and blastula of *Triturus pyrrhogaster*. *Embryologia* **9**, 223–37.

Nakamura, O. & Takasaki, H. (1970). Further studies on the differentiation capacity of the dorsal marginal zone in the morula of *Triturus pyrrhogaster*. *Proc. Jap. Acad.* **46**, 546–51.

Needham, A. E. (1965). Regeneration in the arthropods and its endocrine control. In *Regeneration in Animals and Related Problems*, ed. V. Kiortsis & H. A. L. Trampusch, pp. 283–323. North-Holland, Amsterdam.

New, D. A. T. (1966). *The Culture of Vertebrate Embryos*. Academic Press, New York & London.

Newrock, K. M. & Raff, R. A. (1975). Polar lobe specific regulation of translation in embryos of *Ilyanassa obsoleta. Dev. Biol.* **42**, 242–61.

Nicholas, J. S. & Hall, B. V. (1942). Experiments on developing rats. II. The development of isolated blastomeres and fused eggs. *J. Exp. Zool.* **90**, 441–59.

Nicolet, G. (1970). Is the presumptive notochord responsible for somite genesis in the chick? *J. Embryol. Exp. Morphol.* **24**, 467–78.

Nicolet, G. (1971). Avian gastrulation. *Adv. Morphogen.* **9**, 231–62.

Nicolis, G. & Prigogine, I. (1977). *Self Organization in Nonequilibrium Systems*. Wiley, New York.

Nieuwkoop, P. D. (1952a). Activation and organization of the central nervous system in amphibians. I. Induction and activation. *J. Exp. Zool.* **120**, 1–31.

Nieuwkoop, P. D. (1952b). Activation and organization of the central nervous system in amphibians. II. Differentiation and organization. *J. Exp. Zool.* **120**, 33–81.

Nieuwkoop, P. D. (1952c). Activation and organization of the central nervous system in amphibians. III. Synthesis of a new working hypothesis. *J. Exp. Zool.* **120**, 83–108.

Nieuwkoop, P. D. (1971). The formation of the mesoderm in urodelean amphibians. III. The vegetalizing action of the Li ion. *Wilhelm Roux' Arch. Entwicklungsmech. Org.* **166**, 105–23.

Nieuwkoop, P. D. (1973). The 'organisation centre' of the amphibian embryo, its origin, spatial organization and morphogenetic action. *Adv. Morphogen.* **10**, 1–39.

Nieuwkoop, P. D. (1977). Origin and establishment of embryonic polar axes in amphibian development. *Curr. Top. Dev. Biol.* **11**, 115–32.

Nieuwkoop, P. D. & Sutasurya, L. A. (1977). *Primordial Germ Cells in the Chordates*. Cambridge University Press.

Novick, A. & Wiener, M. (1957). Enzyme induction as an all-or-none phenomenon. *Proc. Natl. Acad. Sci. USA* **43**, 553–66.

Nüsslein-Volhard, C. (1977). Genetic analysis of pattern formation in the embryo of *Drosophila melanogaster*. Characterization of the maternal effect mutant *bicaudal. Wilhelm Roux' Arch. Dev. Biol.* **183**, 249–68.

Nüsslein-Volhard, C., Lohs-Schardin, M., Sander, K. & Cremer, C. (1980). A dorsoventral shift of embryonic primordia in a new maternal effect mutant of *Drosophila. Nature (Lond.)* **283**, 474–6.

Nüsslein-Volhard, C. & Wieschaus, E. (1980). Mutations affecting segment number and polarity in *Drosophila. Nature (Lond.)* **287**, 795–801.

Okada, T. S. (1953). Role of the mesoderm in the differentiation of endodermal organs. *Mem. College Sci., Kyoto Univ.* **20**, 157–62.

Okada, T. S. (1957). The pluripotency of the pharyngeal primordium in urodelan neurulae. *J. Embryol. Exp. Morphol.* **5**, 438–48.

Okada, T. S. (1960). Epitheliomesenchymal relationships in the regional differentiation of the digestive tract in the amphibian embryo. *Wilhelm Roux' Arch. Entwicklungsmech. Org.* **152**, 1–21.

Okada, Y. K. & Hama, T. (1945). Prospective fate and inductive capacity of the dorsal lip of the blastopore of the *Triturus* gastrula. *Proc. Imperial Acad. (Tokyo)* **21**, 342–8.

Okazaki, K. (1975). Normal development to metamorphosis. *The Sea Urchin Embryo*, ed. G. Czihak, pp. 177–232. Springer Verlag, Berlin & Heidelberg.

Ortolani, G. (1955). The presumptive territory of the mesoderm in the ascidian germ. *Experientia* **11**, 445–6.

Othmer, H. G. & Pate, E. (1980). Scale invariance in reaction diffusion models of spatial pattern formation. *Proc. Natl. Acad. Sci. USA* **77**, 4180–4.

Papaioannou, V. E. (1982). Lineage analysis of inner cell mass and trophectoderm using microsurgically reconstituted mouse blastocysts. *J. Embryol. Exp. Morphol.* **68**, 199–209.

Pasteels, J. (1938). Recherches sur les facteurs initiaux de la morphogenèse chez les amphibiens anoures. *Arch. Biol.* **48**, 629–67.

Pasteels, J. (1942). New observations concerning the maps of presumptive areas of the young amphibian gastrula (*Ambystoma* and *Discoglossus*). *J. Exp. Zool.* **89**, 255–81.

Pavlidis, T. (1973). *Biological Oscillators: Their Mathematical Analysis.* Academic Press, New York & London.

Pearson, M. & Elsdale, T. (1979). Somitogenesis in amphibian embryos. I. Experimental evidence for an interaction between two temporal factors in the specification of somite pattern. *J. Embryol. Exp. Morphol.* **51**, 27–50.

Pedersen, R. A. & Spindle, A. I. (1980). Role of the blastocoel microenvironment in early mouse embryo differentiation. *Nature (Lond.)* **284**, 550–2.

Penners, A. (1926). Experimentelle Untersuchungen zum Determinationsproblem am Keim von *Tubifex rivulorum* Lam. II. Die Entwicklung teilwise abgetöteter Keime. *Z. Wiss. Zool.* **127**, 1–140.

Poulson, D. F. (1950). Histogenesis, organogenesis and differentiation in the embryo of *Drosophila melanogaster.* In *The Biology of Drosophila*, ed. M. Demerec, pp. 168–274. Wiley, New York.

Rau, K. G. & Kalthoff, K. (1980). Complete reversal of anteroposterior polarity in a centrifuged insect embryo. *Nature (Lond.)* **287**, 635–7.

Raven, Ch. P. (1966). *Morphogenesis: The Analysis of Molluscan Development*, 2nd edn. Pergamon Press, New York.

Rawles, M. E. (1936). A study in the localization of the organ forming areas in the chick blastoderm of the head process stage. *J. Exp. Zool.* **32**, 271–315.

Razin, A. & Riggs, A. D. (1980). DNA methylation and gene function. *Science* **210**, 604–10.

Reverberi, G. & Minganti, A. (1946). Fenomeni di evocazione nello sviluppo dell'uovo di Ascidie. *Publ. Staz. Zool. Napoli* **20**, 199–252.

Reverberi, G. & Ortolani, G. (1962). Twin larvae from halves of the same egg in ascidians. *Dev. Biol.* **5**, 84–100.

Reverberi, G. & Ortolani, G. (1963). On the origin of the ciliated plates and of the mesoderm in the Ctenophores. *Acta Embryol. Morphol. Exp.* **6**, 175–90.

Runnstrom, J. (1975). Integrating factors. In *The Sea Urchin Embryo*, ed. G. Czihak, pp. 646–70. Springer Verlag, Berlin & Heidelberg.

Rutter, W. J., Clark, W. R., Kemp, J. D., Bradshaw, W. S., Sanders, T. G. & Bell, W. D. (1968). Multiphasic regulation in cytodifferentiation. In *Epithelial–Mesenchymal Interactions*, ed. R. Fleischmajer & E. Billingham, pp. 114–31. Williams & Wilkins, Baltimore.

Ruud, G. & Spemann, H. (1922). Die Entwicklung isolierter dorsaler und lateraler Gastrulahälften von *Triton taeniatus* und *alpestris*, ihre Regulation und Postgeneration. *Arch. Entwicklungsmech.* **52**, 95–165.

Sander, K. (1960). Analyse des ooplasmatischen Reactionssystems von *Euscelis plebejus* Fall. (Cicadina) durch isolieren und Kombinieren von Keimteilen. II. Die Differenzierungsleistungen nach Verlagern von Hinterpolmaterial. *Wilhelm Roux' Arch. Entwicklungsmech. Org.* **151**, 660–707.

Sander, K. (1976). Specification of the basic body pattern in insect embryogenesis. *Adv. Insect. Physiol.* **12**, 125–238.

Satoh, N. (1979). On the 'clock' mechanism determining the time of tissue specific enzyme development during ascidian embryogenesis. I. Acetylcholinesterase development in cleavage arrested embryos. *J. Embryol. Exp. Morphol.* **54**, 131–9.

Satoh, N. & Ikegami, S. (1981*a*). A definite number of aphidicolin sensitive cell cycle events are required for acetylcholinesterase development in the presumptive muscle cells of the ascidian embryo. *J. Embryol. Exp. Morphol.* **61**, 1–13.

Satoh, N. & Ikegami, S. (1981*b*). On the 'clock' mechanism determining the time of tissue specific enzyme development during ascidian embryogenesis. II. Evidence for association of the clock with the cycle of DNA replication. *J. Embryol. Exp. Morphol.* **64**, 61–71.

Saxén, L. (1961). Transfilter neural induction of amphibian ectoderm. *Dev. Biol.* **3**, 140–52.

Saxén, L. & Toivonen, S. (1962). *Primary Embryonic Induction.* Logos Press, London.

Scharf, S. R. & Gerhart, J. C. (1980). Determination of the dorsoventral axis in eggs of *Xenopus laevis*: complete rescue of UV impaired eggs by oblique orientation before first cleavage. *Dev. Biol.* **79**, 181–98.

Seidel, F. (1929). Untersuchungen über das Bildungsprinzip der Keimanlage im Ei der Libelle *Platycnemis pennipes. Wilhelm Roux' Arch. Entwicklungsmech. Org.* **119**, 322–440.

Seidel, F. (1935). Der Anlageplan im Libellenei. *Wilhelm Roux' Arch. Entwicklungsmech. Org.* **132**, 671–751.

Seidel, F. (1960). Die Entwicklungsfähigkeiten isolierter Furchungszellen aus dem Ei des Kaninchens *Oryctolagus cuniculus. Wilhelm Roux' Arch. Entwicklungsmech. Org.* **152**, 43–130.

Sengel, P. (1976). *The Morphogenesis of Skin.* Cambridge University Press.

Shen, G. (1937). Experimente zur Analyse der Regulationsfähigkeit der frühen Gastrula von *Triton*, zugleich ein Beitrag zum Problem der Cyclopie. *Wilhelm Roux' Arch. Entwicklungsmech. Org.* **137**, 271–316.

Sherman, M. I. (1975). Long term culture of cells derived from mouse blastocytes. *Differentiation* **3**, 51–67.

Škreb, N., Švajger, A. & Levak-Švajger, B. (1976). Developmental potentialities of the germ layers in mammals. In *Embryogenesis in Mammals*, CIBA Symposium 40, pp. 27–45. Elsevier, Amsterdam.

Slack, C. & Warner, A. E. (1973). Intracellular and intercellular potentials in the early amphibian embryo. *J. Physiol. (Lond.)* **232**, 313–30.

Slack, J. M. W. (1980*a*). A serial threshold theory of regeneration. *J. Theoret. Biol.* **82**, 105–40.

Slack, J. M. W. (1980*b*). Regulation and potency in the forelimb rudiment of the axolotl embryo. *J. Embryol. Exp. Morphol.* **57**, 203–17.

Slack, J. M. W. (1982). Regeneration and the second anatomy of animals. In *Developmental Order*, Symposia of the Society for Developmental Biology 40, ed. S. Subtelny & P. B. Green, pp. 423–36. Academic Press, New York & London.

Slack, J. M. W. & Forman, D. (1980). An interaction between dorsal and ventral regions of the marginal zone in early amphibian embryos. *J. Embryol. Exp. Morphol.* **56**, 283–99.

Smith, L. D. (1966). The role of a 'germinal plasm' in the formation of primordial germ cells in *Rana pipiens. Dev. Biol.* **14**, 330–47.

Snell, G. D. & Stevens, L. C. (1966). Early embryology. In *Biology of the Laboratory Mouse*, ed. E. L. Green, pp. 205–45. Dover Publications, New York.

Snow, M. H. L. (1976). Embryo growth during the immediate postimplantation period. In *Embryogenesis in Mammals*, CIBA Symposium 40, pp. 53–70.. Elsevier, Amsterdam.

Sonnenblick, B. P. (1950). The early embryology of *Drosophila melanogaster*. In *The Biology of Drosophila*, ed. M. Demerec, pp. 62–167. Wiley, New York.

Spemann, H. (1931). Über den Anteil von Implantat und Wirtskeim an der Orientierung und Beschaffenheit der induzierten Embryonalanlage. *Wilhelm Roux' Arch. Entwicklungsmech. Org.* **123**, 389–517.

Spemann, H. (1938). *Embryonic Development and Induction.* (Reprinted in 1967 by Hafner, New York.)

Spemann, H. & Mangold, H. (1924). Über Induktion von Embryonenanlagen durch Implantation artfremder Organisatoren. *Arch. microsk. Anat. Entwicklungsmech.* **100**, 599–638.

Spirin, A. S. (1966). On 'masked' forms of messenger RNA in early embryogenesis and in other differentiating systems. *Curr. Top. Dev. Biol.* **1**, 1–38.

Spratt, N. T. & Haas, H. (1960). Integrative mechanisms in the development of the early chick blastoderm. I. Regulative potentiality of separated parts. *J. Exp. Zool.* **145**, 97–137.

Spratt, N. T. & Haas, H. (1961). Integrative mechanisms in the development of the early chick blastoderm. III. Role of cell population size and growth potentiality in synthetic systems larger than normal. *J. Exp. Zool.* **147**, 271–93.

Spratt, N. T. & Haas, H. (1965). Germ layer formation and the role of the primitive streak in the chick. *J. Exp. Zool.* **158**, 9–38.

Steiner, E. (1976). Establishment of compartments in the developing leg imaginal discs of *Drosophila melanogaster. Wilhelm Roux's Arch. Dev. Biol.* **180**, 9–30.

Stent, G. S. & Weisblat, D. A. (1982). The development of a simple nervous system. *Sci. Am.* **246**, 100–11.

Stevens, L. C. (1980). Teratocarcinogenesis and spontaneous parthenogenesis in mice. In *Differentiation and Neoplasia*, pp. 265–74. Springer Verlag, Berlin & Heidelberg.

Strickland, S. & Mahdavi, V. (1978). The induction of differentiation in teratocarcinoma stem cells by retinoic acid. *Cell* **15**, 393–403.

Struhl, G. (1981). A gene product required for correct initiation of segmental determination in *Drosophila. Nature (Lond.)* **293**, 36–41.

Sturtevant, A. H. (1923). Inheritance of direction of coiling in *Limnaea. Science* **58**, 269–70.

Sturtevant, A. H. (1929). The *claret* mutant type of *Drosophila simulans*: a study of chromosome elimination and cell lineage. *Z. Wiss. Zool.* **135**, 323–56.

Sulston, J. E. & Horvitz, H. R. (1977). Postembryonic cell lineages of the nematode *Caenorhabditis elegans. Dev. Biol.,* **56**, 110–56.

Summerbell, D. (1976). A descriptive study of the rate of elongation and differentiation of the skeleton of the developing chick wing. *J. Embryol. Exp. Morphol.* **35**, 241–60.

Summerbell, D., Lewis, J. H. & Wolpert, L. (1973). Positional information in chick limb morphogenesis. *Nature (Lond.)* **224**, 492–6.

Takaya, H. (1978). Dynamics of the organizer. In *The Organizer*, ed. O. Nakamura & S. Toivonen, chapt. 2A. Elsevier/North-Holland, Amsterdam.

Tarin, D., Toivonen, S. & Saxén, L. (1973). Studies on ectodermal–mesodermal relationship in neural induction. II. Intercellular contacts. *J. Anat.* **115**, 147–8.

Tarkowski, A. K. (1959). Experiments on the development of isolated blastomeres of mouse eggs. *Nature (Lond.)* **184**, 1286–7.

Tarkowski, A. K. (1961). Mouse chimeras developed from fused eggs. *Nature (Lond.)* **190**, 857–60.

Tarkowski, A. K. & Wroblewska, J. (1967). Development of blastomeres of mouse eggs isolated at the 4 and 8 cell stage. *J. Embryol. Exp. Morphol.* **18**, 155–80.

Tazima, Y. (1964). *The Genetics of the Silkworm*. Logos Press, London.

Thomas, R. (1973). Boolean formalization of genetic control circuits. *J. Theoret. Biol.* **42**, 563–85.

Tiedemann, H. (1976). Pattern formation in early developmental stages of amphibian embryos. *J. Embryol. Exp. Morphol.* **35**, 437–44.

Titlebaum, A. (1928). Artifical production of Janus embryos of *Chaetopterus. Proc. Natl. Acad. Sci. USA* **14**, 245–7.

Toivonen, S. (1978). Regionalisation of the embryo. In *The Organizer*, ed. O. Nakamura & S. Toivonen, chapt. 2. Elsevier/North-Holland, Amsterdam.

Turing, A. M. (1952). The chemical basis of morphogenesis. *Phil. Trans. R. Soc. B* **237**, 37–72.

Turner, F. R. & Mahowald, A. P. (1976). Scanning electron microscopy of *Drosophila* embryos. I. The structure of the egg envelope and the formation of the cellular blastoderm. *Dev. Biol.* **50**, 95–108.

Tyler, A. (1930). Experimental production of double embryos in molluscs. *J. Exp. Zool.* **57**, 347–407.

Ullman, S. L. (1964). The origin and structure of the mesoderm and the formation of the coelomic sacs in *Tenebrio molitor* L. (Insecta, Coleoptera), *Phil. Trans. R. Soc. B* **248**, 245–77.

Ursprung, H. & Nöthiger, R. (1972). *The Biology of Imaginal Discs.* Results and Problems in Cell Differentiation 5. Springer Verlag, Berlin & Heidelberg.

Vakaet, L. (1962). Some new data concerning the formation of the definitive endoblast in the chick embryo. *J. Embryol. Exp. Morphol.* **10**, 38–57.

Van Blerkom, J., Barton, S. C. & Johnson, M. H. (1976). Molecular differentiation in the preimplantation mouse embryo. *Nature (Lond.)* **259**, 319–21.

van den Biggelaar, J. A. M. (1977). Development of dorsoventral polarity and mesentoblast determination in *Patella vulgata. J. Morphol.* **154**, 157–86.

van den Biggelaar, J. A. M. & Guerrier, P. (1979). Dorsoventral polarity and mesentoblast determination as concomitant results of cellular interactions in the mollusc *Patella vulgata. Dev. Biol.* **68**, 462–71.

Van der Meer, J. M. (1979). The specification of metameric order in the insect *Callosobruchus maculatus* Fabr. (Coleoptera). I. Incomplete segment patterns can result from constriction induced cytological damage to the egg. *J. Embryol. Exp. Morphol.* **51**, 1–26.

Van Dongen, C. A. M. (1976). The development of *Dentalium* with special reference to the significance of the polar lobe. V. Differentiation of the cell pattern in lobeless embryos of *Dentalium vulgare* (da Costa) during late larval development. *Proc. K. Ned. Akad. Wet., Ser. C* **79**, 245–55.

Van Dongen, C. A. M. & Geilenkirchen, W. L. M. (1974). The development of *Dentalium* with special reference to the significance of the polar lobe. I–IV. *Proc. K. Ned. Akad. Wet., Ser. C* **77**, 57–100.

Verdonk, N. H. (1968). The effect of removing the polar lobe in centrifuged eggs of *Dentalium. J. Embryol. Exp. Morphol.* **19**, 33–42.

Verdonk, N. H. & Cather, J. N. (1973). The development of isolated blastomeres in *Bithynia tentaculata. J. Exp. Zool.* **186**, 47–61.

Vogel, O. (1978). Pattern formation in the egg of the leafhopper *Euscelis plebejus* Fall. (Homoptera). Developmental capacities of fragments isolated from the polar egg regions. *Dev. Biol.* **67**, 357–70.

Vogt, W. (1929). Gestaltungsanalyse am Amphibienkeim mit örtlicher Vitalfärbung. II. Gastrulation and Mesodermbildung bei Urodelen und Anuren. *Wilhelm Roux' Arch. Entwicklungsmech. Org.* **120**, 384–706.

von Ubisch, L. (1938). Über Keimverschmeltzungen an *Ascidiella aspersa. Wilhelm Roux' Arch. Entwicklungsmech. Org.* **138**, 18–36.

Waddington, C. H. (1932). Experiments on the development of chick and duck embryos, cultivated *in vitro. Phil. Trans. R. Soc. B* **221**, 179–230.

Waddington, C. H. (1933). Induction by the endoderm in birds. *Wilhelm Roux' Arch. Entwicklungsmech. Org.* **128**, 502–21.

Waddington, C. H. (1934). Experiments on embryonic induction. *J. Exp. Biol.* **11**, 211–27.

Waddington, C. H. (1952). *The Epigenetics of Birds.* Cambridge University Press.

Warner, A. E. (1973). The electrical properties of the ectoderm in the amphibian embryo during induction and early development of the nervous system. *J. Physiol. (Lond.)* **235**, 267–86.

West, J. D. (1976). Clonal development of the retinal epithelium in mouse chimeras and X inactivation mosaics. *J. Embryol. Exp. Morphol.* **35**, 445–61.

Whittington, P. Mc.D. & Dixon, K. E. (1975). Quantitative studies of germ plasm and germ cells during early embryogenesis of *Xenopus laevis. J. Embryol. Exp. Morphol.* **33**, 57–74.

Whittaker, J. R. (1973). Segregation during ascidian embryogenesis of egg cytoplasmic information for tissue specific enzyme development. *Proc. Natl. Acad. Sci. USA* **70**, 2096–100.

Whittaker, J. R. (1980). Acetylcholinesterase development in extra cells by changing the distribution of myoplasm in ascidian embryos. *J. Embryol. Exp. Morphol.* **55**, 343–54.

Whittaker, J. R., Ortolani, G. & Farinella-Ferruzza, N. (1977). Autonomy of acetylcholinesterase differentiation in muscle lineage cells of ascidian embryos. *Dev. Biol.* **55**, 196–200.

Wieschaus, E. & Gehring, W. (1976a). Clonal analysis of primordial disc cells in the early embryo of *Drosophila melanogaster. Dev. Biol.* **50**, 249–65.

Wieschaus, E. & Gehring, W. (1976b). Gynandromorph analysis of the thoracic disc primordia in *Drosophila melanogaster. Wilhelm Roux's Arch. Dev. Biol.* **180**, 31–46.

Wilson, E. B. (1904a). Experimental studies on germinal localization. I. The germ regions in the egg of *Dentalium. J. Exp. Zool.* **1**, 1–72.

Wilson, E. B. (1904b). Experimental studies on germinal localization. II. Experiments on the cleavage mosaic in *Patella* and *Dentalium. J. Exp. Zool.* **1**, 197–268.

Wilson, E. B. (1904c). Mosaic development in the annelid egg. *Science* **20**, 748–50.

Wilson, E. B. (1929). The development of egg fragments in annelids. *Wilhelm Roux' Arch. Entwicklungsmech. Org.* **117**, 179–210.

Winfree, A. T. (1980). *The Geometry of Biological Time.* Springer Verlag, New York.

Wolpert, L. (1969). Positional information and the spatial pattern of cellular differentiation. *J. Theor. Biol.* **25**, 1–47.

Wolpert, L. (1971). Positional information and pattern formation. *Curr. Top. Dev. Biol.* **6**, 183–224.

Woo Youn, B. & Malacinski, G. M. (1980). Action spectrum for ultraviolet irradiation inactivation of a cytoplasmic component required for neural induction in the amphibian egg. *J. Exp. Zool.* **211**, 369–77.

Wright, D. A. & Lawrence, P. A. (1981). Regeneration of the segment boundary of *Oncopeltus. Dev. Biol.* **85**, 317–27.

Yagil, C. & Yagil, E. (1971). On the relation between enzyme concentration and the rate of induced enzyme synthesis. *Biophys. J.* **11**, 11–27.

Yajima, H. (1960). Studies on embryonic determination of the harlequin fly *Chironomus dorsalis.* I. Effects of centrifugation and of its combination with constriction and puncturing. *J. Embryol. Exp. Morphol.* **12**, 89–100.

Yamada, T. (1937). Der Determinationzustand des Rumpfmesoderms in Molchkeim nach der Gastrulation. *Wilhelm Roux' Arch. Entwicklungsmech. Org.* **137**, 151–270.

Yamada, T. (1967). Cellular and subcellular events in Wolffian lens regeneration. *Curr. Top. Dev. Biol.* **2**, 247–83.

Yatsu, N. (1911). Observations and experiments on the ctenophore egg. II. Notes on the early cleavage stages and experiments on cleavage. *Annot. Zool. Japon.* **7**, 333–46.

Yatsu, N. (1912). Observations and experiments on the ctenophore egg. III. Experiments on germinal localization of the egg of *Beroe ovata. Annot Zool. Japon.* **8**, 5–13.

Zalokar, M., Erk, I. & Santamaria, P. (1980). Distribution of Ring-X chromosomes in the blastoderm of gynandromorphic *D. melanogaster. Cell* **19**, 133–41.

Ziomek, C. A. & Johnson, M. H. (1980). Cell surface interaction induces polarization of mouse 8 cell blastomeres at compaction. *Cell* **21**, 935–42.

Index